# Sustainable food consumption

# Sustainable food consumption

## A practice based approach

Elizabeth Sargant

*Environmental Policy Series – Volume 11*

EAN: 9789086862634
e-EAN: 9789086868117
ISBN: 978-90-8686-263-4
e-ISBN: 978-90-8686-811-7
DOI: 10.3920/978-90-8686-811-7

First published, 2014


Wageningen Academic Publishers
P.O. Box 220
6700 AE Wageningen
The Netherlands
www.WageningenAcademic.com
copyright@WageningenAcademic.com

Voor mijn opa (1915) die na het bidden voor een maaltijd vaak zei: 'Voor goed eten mag je best danken'

Voor mijn vader (1945-2006) die vaak zei: 'Voor goed eten mag je best wat meer betalen'

# Preface

The natural world and life sciences have always interested me and thus I began my 'academic journey' with a bachelor's degree in ecology and environment biology. In those days I was mostly driven by idealism, a wish to make some contribution to a better world. I spent much of my years as a student learning many interesting things, dabbling in various disciplines, from microbiology, to aquatic ecology and toxicology, to social and political sciences in the area of environmental studies. When I had the opportunity to do a PhD at the Environmental Policy Group (ENP) in Wageningen I was extremely happy, I could not think of a more valuable thing than trying to understand how to connect the large-scale, abstract goal of sustainable development, with the small-scale of everyday-life consumption.

Agricultural and food consumption practices are the most important contributors to ecosystem degradation. The modern agricultural system is responsible for putting more carbon dioxide into the atmosphere than the actual burning of fossil fuel. It degrades the soils needed to grow nutritious foods. Wendell Berry has rightly pointed it out: 'eating is an agricultural act'. Through doing this PhD I have come to realise that more sustainable patterns of food consumption require nothing less than fundamental societal change. On an individual level I think this requires us to be open minded, to be willing to learn. It requires us to become less dogmatic and practice a certain degree of modesty. We cannot move further if we chain ourselves to rigid frameworks, paradigms and assumptions. We need new solutions and we need to be creative.

My career as a PhD student was far from ideal; it had wonderful highs but also some serious lows. It took me nine years to finish my degree! Still, I hope this PhD, or some part of it, might inform and/or inspire someone in some way. At ENP my promotor and colleagues remained patient with me, and always made me feel welcome at the group. I would like to thank all the people at ENP, but especially Sara Statham, Jennifer Lenhart, Dorien Korbee, Judith van Leeuwen, Judith Tukahirwa, Eira Carballo Cardenas, Carolina Maciel, Alice Miller, Dries Hegger and Hilde Toonen. Thank you to Corry Rothuizen for all her help over the years, and all those nice chats in her office. A very special thank you is reserved for my promotors Gert Spaaargaren and Hans Mommaas (Tilburg University) for all their support, input and dedication. Gert, thank you for not giving up on me. Also, I would like to thank all those involved in the CONTRAST project for all their hard work: Eric Drissen, Hans Dagevos, Sander van den Burg, Bertine Bargeman and Loes Maas. Thank you to Astrid Hendriksen for her facilitating my focus groups and helping me with methodological questions. Thanks to my fellow PhD students involved in CONTRAST: Jorrit Nijhuis and Desiree Verbeek, and especially Lenny Putman for all her support and her excellent mind.

A special thank you goes out to Petra Derkzen; our talks always make me feel super motivated. Our small endeavour to improve the catering at the university actually helped me remain enthusiastic about my own research, and even made me understand how powerful it is when you don't just sit around talking about change but try to organise it yourself.

I would like to thank my colleagues at Lazuur Food community in Wageningen for what they have taught me these past years, and for all the time-off they allowed me in order to work on my PhD. Without their support I might not have been able to finish it!

A BIG thank you to my warm and wonderful family for all their love and support, especially my mother Nel and mother-in-law Christien (also thanks for the extra babysitting you both did). Thank you: Opa, Sander, Norah and Lilly; Maaike, Maarten and little Roos; Rose, Valter and Oskar; John and Bernadine. Thank you to Barbara Pescadinha and Charlotte Phillips for your paranymph services and your friendship. And my friends cheering at the side lines: Danielle Phillips, Sara Mulder, Corine van der Heide, Kristian Maters, Frans Bushman and their families.

I owe much to my partner (and fiancé) Sander Janssen for all that he has done for me throughout this PhD. Thank you my love for all your motivational speeches, for proof reading and editing, for EndNote management and editing references, for more motivational speeches, for much love and tenderness and your endless patience and strong, positive spirit. Obviously I could not have done this without you!

# Table of contents

# Chapter 1. Introduction

## 1.1 Introduction

The sustainable development of food and agriculture is linked to just about every aspect of human society. It touches on the most essential global and local problems of our time (LNV, 2009). Thinking about how to make food production and consumption more sustainable is not just a question of how to make agriculture and food production less environmentally unfriendly. It also involves fundamental questions about our relationship with food, with nature and with each other. It involves questions about the definitions of what we consider to be good food and good agriculture and how sustainability should be realised, through increased scale and technology, or a more nature-bound system (Lang & Heasman, 2004). Our everyday food practices sometimes seem far removed from these issues. Although we might be aware of sustainability issues surrounding food and agriculture, the actual implications for own food habits are not always so evident.

The attention paid to promoting sustainable consumption patterns has increased over recent years (DEFRA, 2006; EC, 2004b; Garnett, 2008; PBL, 2008). The Dutch government wishes to create a food and agriculture sector which is more in tune with values pertaining to animal rights, human health and the environment (LNV, 2009; WUR, 2008). Sustainable food products and services are considered one of the fasted growing, stable markets within the food retail and services sector (EZ, 2012). However, various studies report that despite the apparent growth in people's awareness of food related sustainability issues, the attention paid to sustainability in everyday choices for food products and services (e.g. organic food, Fair Trade products, eating less meat) remains marginal (Bartels *et al.*, 2009; WUR, 2008). In order to remedy this, much research is geared toward a better understanding of consumers' individual preferences, values and attitudes in order to adapt 'green' messages, products and services in such a way as to be more appealing. However this is often done with disregard to the contexts and settings consumers find themselves in and the (infra)structures of food provisioning. Indeed with the focus so firmly on individual rationale one is in danger of missing out on the 'bigger picture'.

This study hopes to contribute to understanding this 'bigger picture' by analysing food consumption from a practice perspective, that is, by focussing on food consumption as a practice and thus incorporating the situational and infrastructural dimension of consumption. Studies have indicated that bad food infrastructure or 'obsogenic' environments promote unhealthy eating patterns in society (Dagevos & Munnichs, 2007; Gezondheidsraad, 2003), and that the availability and accessibility of healthy foods is paramount in promoting healthier eating patterns (Brug & Lenthe, 2005; Kamphuis *et al.*, 2006). The same is likely to be applicable to (un)sustainable eating habits. Taking context into account may enable us to see possibilities and opportunities which we might otherwise not. Food consumption takes place at work, at school, in train stations, in petrol stations, etc. What foods are eaten here? Who determines the content and taste of these foods? Why is there so little focus on what is happening in these area's?

This research is interested in the position of the consumer within the sustainable development of the food domain and the ways in which everyday food consumption might improve in terms of sustainability. For the purpose of this research we take the sustainable development of food consumption to mean increasing the use of and engagement in various sustainable food alternatives[1] but also the improvement in provisioning of sustainable alternatives in a boarder range of consumption contexts within society, like within canteen facilities within offices and educational facilities, cafés and restaurants, etc. This study investigates how the transition to more sustainable food consumption in society can be looked at using a practice-based approach. Firstly it considers the position of the consumer with the sustainable development of the agro-food sector. Secondly, it studies the extent to which Dutch consumers engage in sustainable food consumption and how consumers with different food shopping practices differ in terms of this engagement. Lastly, it considers sustainable food provisioning and consumption within a specific context outside the home; public and private canteen catering.

## 1.2 Sustainable development in food provisioning in a nutshell

More than ever consumers have the opportunity to find food products and services which are less environmentally unfriendly, which consider animal and ecological welfare, and famers welfare at home and abroad. A vast array of technologies, agricultural systems, social movements and agro-entrepreneurial activities contribute to creating a more ecologically and socially sound (global) agro-food system. The alternative food sector, notably organic/biodynamic agriculture and retail, has slowly but steadily grown into a professional alternative food provisioning system (Table 1.1).

*Table 1.1. Dutch organic food sales and sale points in 2012 (Bionext, 2012).*

| | **Approx. number of points of sale 2012** | **Approx. return organic food products 2012 (× million euro's)** |
|---|---|---|
| Discounters | 866 | 7 |
| Marqt (food shop) | 5 | 8 |
| Drugstores | 1000 | 1 |
| Butchers | 20 | 1 |
| Internet sales | 130 | 2.5 |
| Markets | 40 | 18 |
| Farm shops | 350 | 17 |
| Vegetable grocers | 1,050 | 1.8 |
| Bakeries | 3,930 | 10 |
| Off licences | 2,800 | 1.8 |
| Cheese and delicacy shops | 1,100 | 1 |

[1] This means the use of either alternative products, like organic food, fair trade products, local and seasonal produce, or practical measures leading to changes in food habits, such as eating less meat, combatting food wastage.

As Marsden has noted, food markets have become more differentiated along a variety of socially constructed food quality criteria (Marsden, 1998). Food provisioning is being 'redefined' along the lines of the concerns of both consumers and producers (Spaargaren *et al.*, 2012). These concerns relate to environmental and (rural) socio-economic issues implicated in industrialised and globalised agro-food systems, but also broader concerns of food quality and safety. Similarly, policymakers and shapers are not only looking at how food quantity should be secured but how agro-food quality should be defined.[2]

There are three major developments or trends which can be used to describe the state of sustainable food provisioning in retail and food services in Western Europe and the Netherlands today.

### 1.2.1 Alternative food networks

In the Netherlands, as in other European countries, the first alternative food networks (AFN's) (Renting *et al.*, 2003; Roep & Wiskerke, 2012) were founded by pioneering farmers and entrepreneurs of the early twentieth century who were concerned about the unviable and unsustainable nature of industrialised farming methods (Winkel, 2012). According to one definition 'AFN's ... [build] new producers-consumer alliances and [create] experimental spaces to develop novel practices of food provision that are more in tune with their values, norms, needs, and desires, that built on the reproduction and revaluation of local sources, and that result in food of distinct and better appreciated qualities' (Roep & Wiskerke, 2012: p. 206). Here we find the renewed attention for aspects such as seasonal produce, crop diversity, produce quality and more healthy and ecologically sound production/processing methods. AFN's are centred around the re-localisation of food chains and the building of ecologically and socially-sound agro-food systems according to principles of ecologically sound resource cycling as well as the link between soil and food quality. AFN's offer a way for producers to attain more stable farm incomes and continuity in the face of globalising food chains and social and ecological degradation of hyper-industrialised and intensive farming (Renting *et al.*, 2003).

From the consumers' perspective the interest in AFN's lies in the search for more trustworthy, better quality foods which are more ecologically and socially sound. In many cases consumers are searching for more authentic, more direct contact with producers of food. Consumers look for alternatives to the 'mass food industry' which has become to hold connotations of low quality, low in taste and unsustainable production. AFN's are created through such initiatives as farmers shops and markets, farmers selling directly to consumers, direct producer-retailer and producer-food service provider cooperations and networks, regional delivery services (Hendriks *et al.*, 2004) and community supported agriculture (CSA, Pergola) initiatives which often include box-schemes (Otters, 2008). In the Netherlands we find various types of CSA initiatives: the Pergola businesses;

[2] Dutch and European policymakers are currently promoting the idea of an agro-food production system based on societal values of agricultural and food quality (Oskam *et al.*, 2005; SER, 2002). The latter includes 'quality' defined in such terms as improved animal welfare, environmental welfare, equitable global trade, countryside community welfare and local food tradition.

adoption projects[3], box-schemes and shops which run according to the 'associative economy' principle where consumers are involved through membership in some form (Otters, 2008). Vegetable and fruit box themes have proven relatively successful in the Netherlands (Melita, 2001).

An important feature of many AFN's is that they are organised by food producers and consumers themselves. All around Europe food producers and consumers 'find each other' in shared needs and wishes for better food. Farmers rely less on the mass-scale logistic and financial system, substituting this with a more direct one-to-one relationship between consumer and producer. Consumers help fund producers through box-schemes and community/consumer funding schemes. Sometimes consumers directly help fund farmers which are starting a farm. Thus, in many AFN's the interests of consumers and producers are aligned, and represent an independent, community-based solution to local/regional socio-economic issues related to food. Arguably, in such instances the AFN's go beyond what labels might offer consumers. In other words, where labels (e.g. EU Eco label EU) still represent more abstract agro-food systems and their institutions, AFN's provide consumers with more personal and tangible access to their daily food produce, and to the famers and the land that produces them.

### *1.2.2 Diversification in food retail and services*

Secondly, there has been a trend towards greater diversification within the food retail and services sector, and an overall professionalization or mainstreaming of the 'alternative' sector. At Biofach 2008 this was identified as one of the major trends in the organic market. 'Large organic supermarkets, organic convenience stores in train stations or shopping centres, organic discounters, organic wine or tea shops, organic butchers and bakers, natural cosmetic and textile shops'.[4] Indeed specialised businesses have appeared on the Dutch market offering organic and Fair Trade snacks in train stations, organic catering and various restaurants that now cook with seasonal, local and/ or organic food. Hollander (2005) notes that after the 'price wars'[5] of Dutch supermarkets in 2003 food retailers had to choose between specialisation strategies focussing on quality, or strategies with the focus on price, a trend also seen in the organic food sector. Within the Netherlands and the UK various shop formulas have been set up (e.g. Marqt, Rio de Bio, Estafette, Goodyfood, Lazuur, the People's Supermarket). These formula's mix different sustainability aspects like animal, environmental and trade welfare, to aspects such as the closer connection with natural processes, (re)localisation, seasonality and rural welfare, as well culinary quality ('good food') and dietary health. In addition we see the appearance of large organic supermarkets in the larger Dutch cities.

[3] The two most well-known of these projects are the adopt a chicken and adopt an apple tree projects. Here one pays a fee for the maintenance of a chicken or an apple tree by a farmer and in return one receives eggs from their own chicken and in the case of the tree, harvest one's own apples.

[4] Http://orgprints.org/13190/01/vaclavik-2008-organicmarket.pdf. Also, for Dutch businesses involved in 'green' food services see website www.puuruiteten.nl.

[5] In the Netherlands a period of extreme competition between Dutch supermarkets, where shop prices drop considerably, is called a 'price war' ('prijzenoorlog').

### *1.2.3 Sustainable development in the agro-food industry and public procurement*

Thirdly, there is the move to greater transparency and the application of sustainability criteria within the Dutch agro-food industry and supermarket industry. This includes such activities as integrating sustainability criteria into production methods and logistics, and increasing product transparency through tracing and certification systems and labelling (e.g. MSC, Fair Trade) (Oosterveer, 2005). Corporate initiatives include those which have started greening parts of their supply chain (e.g. Unilever with MSC), those which have 'greened' their entire business/services (e.g. Fishes, Peeze), wholesalers with their own certification schemes (e.g. Eosta) and non-governmental certification organisations working with business in setting up sustainable production and procurement criteria (e.g. the Dutch Milieukeur).

On the part of the Dutch government we see various (past) initiatives such as the Dutch covenants to stimulate the market for organic food products and the Platform Sustainable Development of Food, both initiated by the former Ministry of Agriculture (LNV) (now called EZ),[6] as well as the sustainable procurement programmes (Agentschap NL, PIANOo). The latter looks to encourage and facilitate sustainable procurement of mainly the public sector in the Netherlands, but also the private sector. Advice on sustainable procurement is provided on subjects: realising/building sustainable buildings, energy saving, cleaning and maintenance and catering. In practice, this means selecting suppliers according to sustainability criteria, and/or demanding environmental and/or social standards from producers/suppliers. Any organisation or institution which procures products and services can be seen as a consumer. The government and public sector has come to be explicit about its responsibility toward sustainable procurement, expressing that it is to be held accountable for its 'consuming behaviour'. Furthermore, it is the protector of public interests, with a clear responsibility to set a good example in terms of acting with responsible consumerism.[7]

## 1.3 Sustainable development in food consumption

### *1.3.1 Consumers and sustainable development*

Consumers and their (potential) actions are given a central role within the realisation of (further) sustainable development of agro-food systems. Such rhetoric fits the general trend to move from central government steering to multi-level forms of 'governance' (Spaargaren & Oosterveer, 2010). Both Dutch and European policymakers wish consumers to 'take their responsibility' and choose more sustainable goods and services. Consumers are considered important 'change agents' within sustainable development. It is often pointed out that consumer acceptance of, and consumer demand for, sustainable products and services are crucial in realising sustainable development

[6] Five major branch organisations representing the food chain (agriculture, industry, retail and food services sector) have agreed to work together under the Dutch Platform Sustainable Development of Food to make improvement in terms of various goals, e.g. energy and water saving, animal welfare, food wastage.

[7] In 2005 the Dutch parliament accepts a motion (Koopmans/De Krom) to commit the public sector to give a heavy weight to sustainability criteria in all its purchasing processes (100% sustainable procurement by 2010).

of agriculture and food (LNV, 2009; SER, 2002). At the same time the ministry of agriculture would like the services which agriculture provides to society to be recognised and rewarded (Houtskoolschets Europees Landbouwbeleid 2020; LNV, 2008). The latter is to be financed through public spending as the market is seen to be unable to adequately provide for the protection of public goods such as nature, the environment, animal welfare and the countryside. It remains unclear how Dutch agro-food policy is to make market prices reflect the negative externalities of unsustainable agro food practices (e.g. damage to ecological systems) (see for example Van Drunen (2010) and how 'food culture' (consumption) is to adapt in such a way as to be more considerate to such aspects as agricultures' 'societal services'. How will they move food producers, providers and consumers to value the links between farming, ecology and social welfare, or, more importantly, get them to 'choose for' agro-food quality (i.e. environmental integrity, 'Fair Trade', animal welfare, etc.)?

Economic and psychological theories approach this question with the premise that food consumption is the result of individuals' functional and/or emotional needs, tastes and preferences. Often the next step in this line of logic is that when consumer behaviour changes in the direction of more sustainable choice, producers will respond accordingly by providing more sustainable products and services. Thus, sustainable consumption needs to be marketed in adequate ways in order to convince the consumer to choose the 'more sustainable' option, thus rejecting the (often cheaper) less sustainable product/service. 'Most policy is based on an understanding that individuals are alert, consciously choosing what they want and what they wish to do, in the light of their values and attitudes ... This model, dominant among educated Europeans and all social sciences since the mid-20$^{th}$ century ... parallels and sustains the model of the sovereign consumer basic to the legitimisation of market exchange and, increasingly, of political action' (Warde, 2013b). Indeed individualistic approaches are likely to have been the status quo for so long within science, policy and business because they bring forth relatively easy realisable solutions such as adapting marketing messages and information provisioning. They fit perfectly in the detracting role of government in consumption and production spheres, transferring responsibility to consumers, and fitting views on consumer sovereignty (Warde, 2013a, 2013b).

However, a whole body of research reveals the social and physical dependencies of consumption (Blake, 1999; Burgess *et al.*, 1998; Dagevos & Munnichs, 2007; Halkier, 2001; Hobson, 2003; Jackson, 2007; Macnaghten, 2003; Shove & Pantzar, 2005; Spaargaren, 1997; Spaargaren *et al.*, 2007a; Warde, 2004b) indicating that there is more than the one-way street between the individual psyche and consumption behaviour. In order to include the interplay between the structural and personal dimensions of consumption, various authors have applied practice theory to studying consumption (Hargreaves, 2008; Shove, 2003; Spaargaren *et al.*, 2007a; Spaargaren & Oosterveer, 2010; Warde, 2005). Practice-based approaches take the social and physical dependencies of consumption into consideration, acknowledging the social and collective nature of food consumption. They investigate the contexts in which consumption takes place, meaning the social and physical infrastructures of consumption, and the actors which have the most (potential) influence on determining consumption practices. For the study of sustainable development of food consumption within society this makes sense because of the following reasons.

### 1.3.2 Context matters

Today, food is eaten and appropriated on a daily basis in various different places and contexts, from the caterer at work, the sandwich bar at the train station or the petrol station, to the Chinese take-away, etc. Food consumption thus involves a whole pallet of different home and out-of-home food consumption practices, each with their own dynamics of food provisioning, modes of access and use, and social interactions. Thus, especially in (sub)urban areas the a major characteristic of food consumption is differentiation (Cheng *et al.*, 2007) and 'flexibility' in terms of access to various food services and the use of convenient 'food solutions' to fit busy lifestyles, tight budgets, etc. and/or possibly a lack of cooking time and skills. Thus, although our food habits have become more individualistic, fitting our individual ideas, preferences and schedules, those with a big hand in shaping food values, preferences and tastes are likely to be food providers in the catering and food retail industry.

Practice perspectives hold that consumption patterns or behaviours, consumer attitudes and opinions on food are shaped by and within these everyday food practices (Warde, 2005). They indicate how different sets of considerations and priorities may play a role in different contexts, how people engage in food consumption for building and maintaining their lives (Jackson, 2007) and practices may be seen as ways to manage food needs in a manner which is personally convenient (Clarke *et al.*, 2006; Connors *et al.*, 2001). These (everyday) meaningful practices can show a certain pattern of choice, which can be seen to be 'rigid' and routinized to different degrees, but never without agency or the possibility for change.

Taking a more contextual perspective on sustainable development of food consumption is also important because aside from unsustainable food practices, unhealthy food consumption is a major problem in society today. In the Netherlands obesity is a growing problem under adults and children and less healthy diets are seen especially within the lower socio-economic classes (Jansen *et al.*, 2002; Kreijl, 2004). Here the lack in fresh vegetables and fruit intake and the excess intake of high-energy foods are two major causes of dietary deficiency (Kreijl, 2004; WHO, 2002).[8] Studies have indicated that our 'obsogenic' environment promotes unhealthy eating patterns (Dagevos & Munnichs, 2007; Gezondheidsraad, 2003), and that the availability and accessibility of healthy foods is paramount in promoting healthier eating patterns (Brug & Lenthe, 2005; Kamphuis *et al.*, 2006). Indeed, the great demand for convenient-to-use food products and meal-components both within households and in the food services and catering industry (Bava *et al.*, 2008; Carrigan *et al.*, 2006; Lang, 2004; Regmi *et al.*, 2005), means less food is made with fresh ingredients and often contains too high levels of added salt and sugar (FNLI, 2008; Gezondheidsraad, 2006).[9] Issues of food sustainability and dietary health are connected, although they are not often approached as such. By employing a perspective which takes context into consideration the two are more likely to meet and opportunities might arise to tackle the two together.

[8] Only 2% of adults between 18 and 30 years old eat a minimum of 150 grams of vegetables per day, and none eat the daily recommended 200 grams (RIVM, 2010).

[9] Also, consumption of food outside the home has also been linked to high energy and fat intake (Kamphuis *et al.*, 2006)

Contexts of food consumption bring forth certain consumption patterns and are 'sites' of meaning creation related to food. Thus, people's associations on food sustainability, safety and quality, and definitions of 'good food' are reproduced and/or generated by participation in food practices. A practice perspective offers the inclusion of context in studying consumption and context is seen as both constraining and enabling. It matters not just because of its negative influencing power on our food habits (e.g. 'obsogenic environments'), but also because of its positive influencing power and the possibilities which it opens up for influencing food habit through culinary appreciation and enjoyment.

## 1.4 Research set-up

The framework used to organize the empirical research in this research is the social practice approach (SPA) (Spaargaren, 1997; Spaargaren *et al.*, 2007a; Spaargaren & Van Vliet, 2000). It merges elements of sociological, cultural and anthropological work on consumption into an approach which puts everyday social practice (e.g. shopping for food, dining out) at the centre of analysis. The social practices approach (SPA) considers consumption in terms of its practical, contextual, everyday nature, making room for both agency and structure. This is done through a further elaboration and specification of the concept of practice(s)' into recognisable a set of everyday (food) consumption practices characterised by corresponding settings or locales for interaction and system(s) of provision (Chapter 2). In this way consumers are conceptualised as practitioners with certain levels of know-how, skills and motivations which are shaped not only by the life history but also through the participation in particular sets of (food) practices (Spaargaren *et al.*, 2007a).

### *1.4.1 Research aims and questions*

The central aim of this research is to employ a more contextual, practice based perspective on studying sustainable food consumption in the Netherlands. There are two specific aims. The first is to discuss the role of the consumer in the sustainable development of food consumption. Shifting from an individualistic to a more context-related perspective on food consumption also implies that the role of the consumer as a 'change agent' and (co)responsible actor must be reconsidered. The second aim is to explore and discuss various routes and strategies for sustainable transitions in food consumption based on a practice approach.

The primary research questions are:

- What insights does a practice based approach generate with regard to the development of more sustainable food consumption practices?
- What kind of strategies for the transition to more sustainable food consumption practices can be derived from studying food consumption using a practice-based approach?

In order to give an answer to these questions desk research and two empirical studies were carried out. The literature study (Chapter 3) is geared at understanding the nature of food consumption today and the position of the consumer within the debate on sustainable food consumption. It discusses the need for a practice-based approach in view of a number of developments in food consumption today. The empirical chapters are introduced below.

### *1.4.2 Research topics of the empirical studies*

The first empirical study (Chapter 4) comprises of a survey investigating the state of Dutch consumers' engagement in sustainable food consumption. Many surveys on sustainable food consumption concentrate on consumer concerns/food values (e.g. animal welfare, environmental and social welfare), attitudes and ('life') values, in relation to only a limited set of sustainable behaviours/alternatives (e.g. meat consumption, organic food). This survey uses SPA to investigate Dutch consumers' engagement in sustainable food consumption. The latter includes views held on consumer co-responsibility, their concerns and perceived attractiveness of various sustainable food alternatives (e.g. eating less meat, wasting less food, organic and Fair Trade products), their daily use of these alternatives and the their evaluation of the availability, accessibility and quality of these alternatives. It also investigates to what degree these aspects differ between consumers with different shopping practices.

The second empirical study (Chapters 5 and 6) investigates food consumption within study and work related settings (higher education, public and private organisations) using qualitative research methods (semi-structured interviews, focus groups). When we look at sociological research, eating 'outside the home' in various contexts (school, work, higher-education, sports centres, etc.) has not been sufficiently considered in relation to sustainable food practices. Research within Europe has focussed mainly on sustainable food procurement policy procurement, with studies looking at sustainable food procurement for schools (Morgan & Sonnino, 2007; Sonnino, 2009; Sustain, 2005). Cases focusing on the Netherlands are few and even less look at the practice of eating in a canteen, something which involves the dynamics between canteen provisioning and the end-user, and the locale, the site where the two meet.

Compared to the first empirical chapter the focus here is less on the individual and more on the interaction between consumers and catering providers as well as what is occurring in daily lunch practices within canteens. Canteen food practices provide interesting and a relatively unexplored subject material. Out-of-home food services have an important role to play in consumers' daily access to healthy and sustainable foods, yet they have not often been the subject of academic inquiry in relation to the topic of sustainable consumerism. Studying this topic means gaining insight into the manner in which sustainable development is occurring in the business catering sector, the promising strategies and bottlenecks involved in this area. This topic is researched.

### *1.4.3 Thesis outline*

Chapter 2 presents the theoretical framework and the research methodology. It provides an introduction into practice theory and discusses how practice theory can be applied to studying sustainable food consumption. It also discusses how the SPA framework is applied and operationalised for research on food consumption, and it introduces the research methodologies used.

Chapter 3 consists of a literature study aimed to discuss some major features of today's food consumption practices and their implications for the role of consumers in the sustainable development of the agro-food sector. It covers issues like the diversity seen in both food provisioning and food practices and the disconnectedness between consumers and their daily meal. It also discusses how a practice-based approach considers encouraging sustainable food consumption.

Chapter 4 presents and discusses the results of the survey findings. The aim of the survey is to explore using a practice-based approach in a quantitative consumer survey, to gain insight into Dutch consumer's engagement in different sustainable food alternatives and the manner in which consumer with different grocery-shopping practices differ in terms of this engagement.

Chapter 5 and 6 present and discusses the study on canteen food consumption. Here the focus lies on gaining a more in-depth understanding of sustainable food consumption in an everyday, out-of-home food practice. Due attention is paid to what is happening within the locales of food consumption and the role of food providers directly involved in shaping everyday consumption practice. Understanding is gained on the (strategic) relevance of considering sustainable food consumption in the catering sector (or, the 'public plate').

Chapter 7 discusses the most important conclusions and gives some recommendations for further research.

# Chapter 2.
# Theory and methodology: a social practice approach to food consumption

## 2.1 Introduction

The manner in which consumption behaviour is studied and framed has important implications for the types of policies considered promising for targeting consumption patterns for sustainable development. In the past much research has been geared toward understanding the link between personal values (such as honesty, idealism, benevolence, etc.) and consumer behaviour. However, various authors have argued and demonstrated that useful insights can be gained from shifting the focus to the contexts of consumption practice(s) (Halkier, 1999, 2009a; Hargreaves, 2008; Shove, 2012a; Spaargaren *et al.*, 2007a). This chapter reviews and discusses some of the most important constituents of this approach which forms the basis of the theoretical framework employed in this study.

Section 2.2 provides a brief discussion of the different perspectives and studies on food consumption. Section 2.3 discusses some key aspects of practice theory which for the backbone of the theoretical framework. Section 2.4 shows how practice theory can be applied to the study of food consumption and Section 2.5 then discusses the theoretical framework, the social practice approach (SPA). Section 2.6 closes with a look at the different research area's and methodologies used in this study.

## 2.2 Introducing perspectives on sustainable consumption

Many studies on consumption base their studies on socio-psychological models or cognitive/rationalist approaches (for review of these see Jackson, 2005), notably value/attitude-behaviour models (Barr & Gilg, 2007; Boer *et al.*, 2007; Burgess, 1992; Grunert & Juhl, 1995; Padel, 2005; Vermeir & Verbeke, 2006, 2008; Zanoli & Naspetti, 2002). Here the individual is taken as the starting point in understanding consumption behaviour. The theory of planned behaviour considers values/beliefs and attitudes as the precursors of behaviour and behavioural intention. Certain types of behaviours are linked to certain sets of values or beliefs about life and society (e.g. Schwartz values, Rokeach values) (Rockeach, 1973; Schwartz, 1992). Various studies have linked personal and/or food related values or qualities to the decision making processes related to sustainable product choices and brand choice (e.g. Bartels *et al.*, 2009; Burgess, 1992; Grunert & Juhl, 1995; Thogersen, 2001). Other studies consider the attitudes towards certain products, or consumer awareness (Vermeir & Verbeke, 2008).

However, socio-psychological theories often divert attention away from the social and physical context of behaviour, and more importantly, the interaction between the individual and this context (Shove, 2010; Shove & Warde, 1998; Warde, 1999). The focus lies on why people do what they do explained in terms of how they perceive the world, or, on people's reports of how they perceive the world. Studies have shown that there is a notorious difference between caring for the environment and actually changing ones consumption patterns accordingly in order to lessen ones impact on the environment (Connolly & Prothero, 2008; Halkier, 1999, 2001, 2004; Macnaghten, 2003), a phenomenon called the attitude-behaviour or value-action gap.

Defining peoples' ability to engage in sustainable behaviours in terms of their values and attitudes does not do credit to the realities of everyday consumption, something which is often habitual, directed at attaining certain expected standards or outcomes and dealing with day to day time management for instance, etc. (Connolly & Prothero, 2008; Macnaghten, 2003). Various scholars have illustrated the complexities of everyday food consumption practices by showing that they are the outcome of conflicting concerns and negotiations between everyday considerations (Dagevos, 2005). Also, individuals may do things differently in different contexts, considering certain aspects almost automatically in some practices, whilst ignoring them in others (Spaargaren & Mol, 2008). Even when people are environmentally knowledgeable this must be weighed up against 'day-to-day' realities (Shove, 2003; Spaargaren & Van Vliet, 2000; Warde, 2004b, 2005). Furthermore, consumption takes place in direct and indirect relationship with socio-material systems, which are the physical and social infrastructures of consumption that precondition certain modes of provisioning, access and use (e.g. Hargreaves, 2008; Jacobsen & Dulsrud, 2007; Spaargaren & Van Vliet, 2000).[10]

In using individualistic approaches one is in greater danger to over-simplify and reduce the consumer to either a passive actor who will take up information and adapt behaviour accordingly, or as an individual who is charged with failing to realise his/her 'responsibility' in view of realising more sustainable consumption patterns. This has implications for how policymakers see the role of consumers within sustainable development and the kinds of policies which are considered viable to target unsustainable consumption (Warde, 2005). In their critique on models which keep the focus on individuals, various authors have indicated that a transition towards more sustainable consumption patterns is a social project, i.e. needing changes on different levels of society, involving different actors, practices and institutions (of practices) (Reckwitz, 2002). 'Social-psychological models are strong in stressing the importance of values and beliefs ... [but] weak about the ways in which individual (motives for) actions should be analytically connected to 'wider society' (Spaargaren & Van Vliet, 2000: p. 5).[11] In line with this, Jacobsen & Dulsrud (2007) point out the importance of looking at the role of other actors' strategies which '... deliberately or not, serve to frame consumers, their options, expectations and self-definitions' (p. 473). This framing is done by various actors within the food chain (e.g. food processors, product developers, facility managers, cooks).

The more contextual approach should be able to take these aspects into consideration. Contextual approaches include contributions from sociological, cultural and anthropological theory on consumption and studies of science, technology and innovation and transition theory. They aim to place the behaviour of individuals in certain structural, social or cultural contexts.

[10] Even researchers who employ value/attitude-behaviour models have worked towards including more contextual factors into their research on consumer choices of sustainable food products, e.g. Tanner & Kast (2003) and Vermeir & Verbeke (2006, 2008).

[11] Reckwitz (2002) points out that the danger with a rationalistic view of agency is that it '... has in diverse ways narrowed our understanding of human agency and the social' (: p. 258, also citing (Taylor, 1985[1971]).

They include actor network theory, socio-material/technical infrastructures theories on the development of technology and socio-technical systems/regimes (Geels, 2004; Schwartz Cowan, 1987; Shove, 2010, 2012a; Shove & Southerton, 2000). Practice theory (Giddens, 1984; Reckwitz, 2002; Schatzki, 2001), which has been developed and applied by a number of disciplines, works towards a more developed analytical balance between the individual and structural aspects. Here practices are the unit of investigation not individuals, groups of individuals or individual values (Shove, 2003, 2010; Shove & Pantzar, 2005; Shove & Walker, 2010; Southerton *et al.*, 2004; Spaargaren *et al.*, 2007a; Warde, 2005).

## 2.3 Practice theory: key concepts

Practice theory is not a distinct and coherent theory, but rather a collection of contributions from philosophers, social theorists, cultural theorists and theorists of science and technology who aimed to strike a balance between structure and agency (Schatzki, 1996). We will review some of the key concepts which make up the essence of the theory on practice and which are relevant to the theoretical framework used in this study.

Reckwitz calls practice theory a kind of cultural theory, distinct from the classical economic models of the *Homo economics* and the sociological model of the *Homo sociologicus*. Contributions to the body of literature come from Giddens, Bourdieu, Foucault, Garfinkel, Latour, Lyotard, Taylor and Schatzki (Eriksen & Nielsen, 2001: p. 130). Bourdieu and Giddens are amongst the early practice theory contributors giving us concepts which prove vital for our understanding of (everyday) practices. Schatzki and Reckwitz are amongst the later contributors.[12]

### *2.3.1 Habitus*

Bourdieu theorised the processes through which individuals and their practices are products of their history and their environment through his concept of 'habitus'. Habitus captures 'the permanent internalisation of the social order in the human body' (Warde, 2004a). A person's habitus can be defined as sets of perceptions, thoughts and actions which are acquired over time resulting from a person's environment (class, family, education). 'The structures constitutive of a particular type of environment ... produce habitus, systems of durable, transposable dispositions, structured structures predisposed to function as structuring structures, that is, as principles of the generation and structuring of practices' (Jackson, 2007). The habitus provides the practical skills needed to participate in different areas of life (fields) (e.g. the arts, different professions) (p. 72). With the concept of habitus, individuals are seen in the context of their social situation and the rules and strategies they employ to engage in a number of everyday practices within different fields of social life. Warde explains habitus as follows: 'The properties of habitus are expressed in terms of how it generates, with respect to many varied areas of practice, schemes of action and the dispositions that generate meaningful practices and meaningful perceptions' (Jackson, 2007: p. 10).

[12] Schatzki produced a solid account on practice theory using a Wittgensteinian approach, whilst Reckwitz showed the position of practice theory as one in a group of cultural theories and showed its implication for a number of social theoretical concepts (e.g. structure, agency, knowledge).

The link between habitus and people's practices is important because it gives practices a historical and contextual dimension, i.e. the way one does things is dependent on one's habitus as generated and reproduced over time. At the same time this concept is not deterministic in that it defines habitus as something resulting from structure alone. Habitus has a 'generative' quality (Bourdieu, 1977; as cited in Jackson, 2007) and as Giddens (1991: p. 81) notes this means that the concept allows one to consider practices and the lifestyles they constitute as processes, 'the result of the continual interplay between agency and the objective condition encountered and experienced' (Giddens, 1984: p. 374).

### *2.3.2 Practice(s) and lifestyle*

Whereas one might say that for Bourdieu the influence of habitus forms the major basis of practices, Giddens sees practices as constituent of an identifiable lifestyle. Giddens defines lifestyle as being made up of 'a more or less integrated set of practices ... not only because such practices fulfil utilitarian needs, but because they give material form to a particular narrative of self-identity' (Giddens, 1991: p. 81). For Giddens individuals cannot be seen separate from both 'internal' and 'external' structures. These structures, or sets of rules and resources, are reproduced in and through practices, as well as being the condition of practices (Giddens, 1984: p. 118). Within this context individual action can be seen to contribute to the creation and maintenance of lifestyles, the locations and vehicles for identity formation (Giddens, 1991).

In addition to structuration theory and the concept of lifestyle, Giddens' concept of 'locale' is important for understanding practices. Locales are 'the use of space to provide the settings of interaction, the settings of interaction in turn being essential to specify its contextuality' (Schatzki, 1996). The concept of locale gives practices a setting which in itself may have certain properties which influence action. Social practices cannot be seen separate from where and when they take place. The 'locale' has certain 'properties specified by the modes of its utilisation (p. 118) and these properties or features of settings 'are also used in a routine manner to constitute the meaningful content of interaction' (p. 119). These features are of a physical nature, comprising of certain artefacts and/or technical and spatial infrastructures. Alternatively, transitions in social practices always imply changes in spatial structures and technical infrastructure as well as changes in social relationships (Reckwitz, 2002: pp. 249-250).

Schatzki argued that individuals, their actions and thoughts cannot be understood independently of the social practices in which they are situated. He further refined practice theory by giving us various notions of practices, two of them being the integrative and the dispersed practices (Schatzki, 1996: p. 99; Warde, 2005: p. 134). The latter are practices of performance such as activities of knowing, learning and understanding (Reckwitz, 2002: p. 256). Integrative practice on the other hand are practices like cooking, travelling, and recreational practices for instance (p. 98). These are described as coordinated entities of 'doings' and 'sayings', which means that analysis of practices must look at both activity and the representation of activity (p. 134). These doings and sayings hang together by means of acceptable ways of 'understandings', rules, principles and instructions ('procedures') and teleoaffective structures ('engagements') (Warde, 2005). In other words, through these components doings and sayings are coordinated within a practice, constituting the very characteristic of that practice.[13] Thus, practices may be described by certain characteristics which people can identify with, and allow them to recognise the practice as such. For instance, the practice of supermarket shopping has some doings and sayings which many of us are familiar with. It involves using shopping carts, looking for the best buys, knowing which shelves contain what kind of products and knowing how to use the technologies in place in some of the supermarkets of today.

### 2.3.3 *Practice(s) and the practitioner*

All three authors contribute to describing and understanding individual actions and behaviour in contextual terms. Such a view then leads to the understanding that people have practice specific, or practice-generated, preferences, ideas, skills and knowledge. Individuals can be seen as practitioners, who have a certain capacity to act and engage in the doings and sayings of practices. Reckwitz explains carefully how to conceptualise knowledge from a practice perspective. Practitioners or agents 'understand the world themselves, and use know-how and motivational knowledge, according to the particular practice' (Reckwitz, 2002: p. 250). 'A specific social practice contains a specific form of knowledge ... Practice embraces ways of understanding, knowing how, ways of wanting and of feeling that are linked to each other within a practice' (p. 253). This also implies that compared to others, an individual may be more or less familiar or competent in the doing of a practice (Hargreaves, 2008; Spaargaren *et al.*, 2007a; Verbeek, 2009), and that the practice may be more or less routinized or habitual to an individual. With this in mind, maybe the most complete definition of a social practice is the one given by Reckwitz: 'a practice is ... a 'type' of behaving and understanding that appears at different locales and at different points of time and is carried out by different bodies/minds' (Clarke *et al.*, 2006; Halkier, 2009a, 2009b; Jackson *et al.*, 2006; Krom, 2008).

## 2.4 Practice theory applied to food consumption

Practice theory offers a focus on human behaviour/action embedded within social and physical contexts, together with the idea that individuals can be seen as knowledgeable practitioners with self-esteem and agency. In a practice based approach, the 'nature and process of consumption' are explained by someone's engagement in practice(s), instead of purely by relying on individually motivated conduct (Warde, 2005: p. 138).

In the last two decades practice theory has been more explicitly applied in empirical and theoretical work on everyday consumption. Authors such as Warde, Shove and Spaargaren have shown how to apply practice theory to the study of consumption. Others have shown how to use practice theory in empirical studies on sustainable consumption behaviour (Hargreaves, 2008; Owens, 2000; Verbeek, 2009). Studies on food consumption which have used practice theory are growing in numbers (Bouwman, 2009; Clarke *et al.*, 2006; Halkier, 2009a, 2009b; Van Otterloo,

[13] Though integrative practice may include dispersed practices, it is more useful from an analytical perspective to see them as separate. Integrative practices are complexes of actors, artefacts and structures, whereas dispersed practices might be seen as involving primarily the minds of people (Warde, 2005).

2000; Van Otterloo & de la Bruhèze, 2002; Van Otterloo & Sluyer, 2000). For instance, Clarke *et al.* (2006) studied how people make use of the shops in their area, looking at consumer choice between and within shops. Halkier (2009b) investigated cooking practices of Danish women and identified different ideal-type cooking styles, focussing on which types of meals are made, how they are made and why.

Applied to food practice theory provides the following insights. Firstly, the use of a practice perspective means one considers sustainable consumption in the social and physical context of social practices. The analytical view is widened from the behaviour of individuals towards an integrated approach on the dynamics and workings of practices. This not only involves analysing what is actually occurring, but also where and when this occurring is made to happen. For instance, considering the practice of eating in a canteen at work not only involves looking at 'individuals' and 'systems', but also at the dynamics occurring within the canteen locale. Questions arise as to the social (with and/or for whom?) and physical (what, where and how) contexts of consumption, and what the role and functioning of various actors is in this particular practice. The focus is on aspects such as provisioning dynamics (who is 'serving the practice' and in what particular manners), access and use patterns (what products are appropriated by participants in what manner), and availability of products and services within certain 'consumption infrastructures' and at specific consumption junctions. Opportunities and limitations for the development of sustainable food practices can thus be identified in the wider context of practices and the situations/settings in which they take place. Indeed a practice perspective encourages us to (and helps us to) consider food consumption in contexts other than the home, for example hospitals, schools, 'food on the go', canteen food, dining out and other food-practices organized at particular locales or consumption junctions.

Secondly, a practice based approach suggests that people's choices, routines, preferences and tastes regarding food are generated by/within, and put into practice within, social practices. As Reckwitz (2002) puts it 'we learn to be bodies in a certain way, in a sense practices are inscribed in bodies, and bodies are vehicles to skilfully carry out practices'. This 'learned' or 'knowledgeable' body is what Bourdieu refers to as habitus, which is built by past practices and thus affects future practices, but does not necessarily have any strategic intention of conduct (e.g. Michael Pollan (2008) shows how the American diet consists primarily of maize). Practices, or experience(s) with food 'generate' standards, wants, desires, concerns and expectations (Barnes, 2001). Food practices are 'inscribed' in our bodies mentally, emotionally and physically, maybe more materially than other human activities. In the case of food, consumption is not just about practices of 'doing', but also about taking in and absorbing. On a micro-scale this means that foods are the very building blocks of our physical, mental and emotional state (Gabriel & Lang, 1995). Food consumption patterns are revealed by the state of our bodies and our tastes. Thus, changing food patterns not only implies changing how we manage or practice food consumption, but also implies changes in bodily functions like tastes, bodily health and physical appearance (e.g. burning fat to lose weight, getting over bodily dependence on sugar).

Thirdly, a practice perspective sees actors in production-consumption chain (so consumers and producers/providers) are practitioners (Warde, 2005: p. 148). The latter means that individuals have some capacity (skills, know-how) with which they carry out, and participate in practices. This 'practitioner capacity' or 'know-how' may be defined and described in various ways, from the understanding and application of certain technologies (e.g. the microwave), to wider knowledge about food (production), nutritional content, where food comes from or how it is grown, etc. Practitioner capacity may be looked at on a micro, everyday level, of how individuals apply materials, skills and knowledge within practices, to more institutional analysis of what type of skills and knowhow is circulating in certain systems of food provisioning, access and use. When discussing skills and knowledge at the micro level of individuals, the concept of 'portfolio' can be used to refer to the skills, knowledge and experiences relevant and needed for the successful participation in a particular (food) practice (see also Chapter 4) (Spaargaren *et al.*, 2007b).

Warde describes practitioner capacity as something which involves 'knowing how to do something' and succeeding in it, a process that which he terms 'internally generated rewards' of practices. 'Judgements of performance are made internally with respect to goals and aspirations of the practice itself, and proficiency and commitment deliver satisfaction and self-esteem' (Cohen & Taylor, 1992: p. 147). In everyday 'ordinary' food consumption practices often involve such things as keeping to a daily schedule, buying food according to one's budget, knowing how to prepare food or receiving enjoyment and satisfaction in eating it with others. These intrinsic rewards may include the confirmation of one's own (cultural, religious) identity and self-understanding (Spaargaren *et al.*, 2007a), keeping up relationships with others, or simply come in the form of momentary enjoyment and personal escape.[14] Spaargaren *et al.* (2007a) also speak of the empowerment and esteem related to practices. In this way, 'the way in which one does things' also lends some identity to a person, and a sense of self-worth. For example, being a good cook gives self-esteem, be it through processes of distinction, or simply through privately experienced benefits. 'The way one does the practice' may be part of one's food identity, but it need not be conspicuous and/or explicit. Representation and conspicuousness may become more important in certain situations, when having people for dinner for instance. Furthermore it plays a role for those who are more actively engaged in the world of food (or the field of food), as a professional cook or a 'foodie'. Here distinction of food knowledge and skill clearly are important (Spaargaren & Van Vliet, 2000).[15]

Indeed this brings us to a fourth point; a practice perspective sees practices as the site of or generator of meanings surrounding food. According to Shove 'conventions that are often taken to constitute the context of behaviour have no separate existence; rather they are themselves sustained and changed through the ongoing reproduction of social practice' (Shove, 2010). Practice theory 'frees' meanings from individuals and places them in their due context. In other words, meanings are not just in the minds of individuals; individuals are not the sole carriers of meanings connected to consumption. Instead meanings exist, and are reproduced, generated or destroyed within socio-physical complexes. This conceptualisation offers room for dynamism, for positive and negative influence, for the opportunity for change.

[14] This is different from the extrinsic rewards of Veblen and Douglas, and Bourdieu's ideas on field where capacity confers distinction and competitive advantage.

[15] Indeed, Warde notes that competitiveness, strategic conduct and distinction ('external goods') are more associated with professional or expert practitioners, and that there are 'ordinary' practices where symbolism and distinction are less prominent characteristics (Warde, 2004a). This refers to Warde's discussion on Bourdieu's concepts of practice and field.

## 2.5 The social practise approach framework

### *2.5.1 Consumption as a social practice*

In the previous two sections we reviewed the most important and relevant aspects of practice theory and considered what it means to study food consumption using a practice based approach. In this section we discuss the social practice approach (SPA) as a conceptual framework for studying everyday food consumption. SPA has its roots in structuration theory of Giddens, and combines this with some key insights from ecological modernisation theory (Mol, 2000). Instead of seeing consumption as a part of or a phase within a practice, consumption itself is seen as and in terms of practices. This is essential in order to conceptually 'connect' consumption to social and physical realities/contexts. Within this research the type of consumption under consideration is 'ordinary', every day and habitual consumption, and food practices have been delineated using the following rule of thumb. A social practice of food consumption it is broadly recognised as a distinct activity, involving distinctive socio-material systems and ways of 'doings' and 'sayings'. Having a canteen lunch for example, is a characteristic food practice, with a certain set up of food provisioning and use-activities which we can all recognise as being 'canteen lunch' (e.g. the type of food provisioning, eating together with colleagues or fellow students). Food consumption practices can consist of appropriation activities (buying food) and/or use(r) activities (preparation, eating, disposal) which may or may not occur within the same locale.

A number of distinct, broadly recognisable integrated social practices social can be defined (e.g. Grocery shopping, eating in a canteen at work, dining, eating 'on-the-go') and if required can be subdivided into sub-practices (Figure 2.1). In theory, each social practice has a corresponding system of provision, a particular locale and involves particular portfolios and lifestyle aspects for its participants. In the social practice of having a canteen lunch in the work place for example, eating in a canteen implies that the consumers or participants to the practice interact with a certain type of provisioning systems, i.e. certain kinds of foods, servants, catering services and formats and particular strategies for sustainable provisioning. At the same time, practitioners or participants to the practice can be characterized for their habitus or lifestyles, as they are shaped by and help shape the behaviour that comes along with participating in the particular practice. Consumers possess prior knowledge and experiences relating to eating in a canteen when entering the practice. They have certain concerns, standards and frames of references which are specific to the 'canteen context' (e.g. on the type of food served there, on the setting of interaction, on the performance of the actors within the system of provision, etc.) and which may differ from other food practices in which they participate over time.

### *2.5.2 Defining practices*

Some additional comments should be made on the formulation of consumption practices for research purposes and features which should be kept in mind when using the practice approach as an analytical tool.

Firstly, researchers should to some extent decide for themselves how to delineate their practices as research objects. The way in which a social practice is delineated should be in service of the

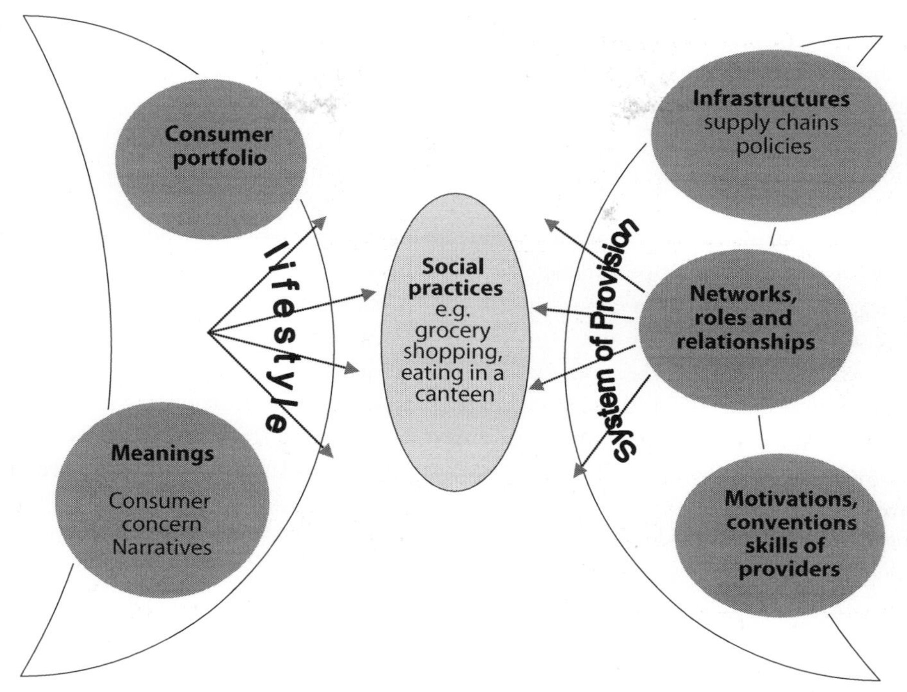


*Figure 2.1. The practice framework (Spaargaren* et al.*, 2007a).*

subject matter, or consumption domain, or case which is being studied. For example, canteen lunch within a work-place setting is one of the practices under scrutiny in this research. However, a different delineation could have been 'having lunch at work'. The first focuses more on canteens whereas the latter takes a broader focus and would include all kinds of lunch-time activities (e.g. going to eat out in a café in town). Furthermore, in some instances it may be useful to speak about sub-practices. For example, eating-out can include fine dining, casual dining and 'food-on-the-go'. One could for instance argue that eating fast food is a practice in itself, and so is eating in a Michelin star restaurant, each with their distinct rules of how to behave and what kinds of foods can be expected. Most important is that food consumption practices are characterised by certain 'rules' (or 'doings' and 'sayings' according to Warde) which are recognisable for both providers and consumers.

Secondly, delineating a social practice of food consumption does not presuppose homogeneity within these practices. For instance, in considering cooking practices one may cover cooking from scratch and using ready-made meals which make for different kinds of meal-time experiences. So practices are never one-of-a-kind, even when discussed under the same banner.

Thirdly, different (food)practices are interconnected to a certain extent, and these connections are made possible by individuals or human agents. Peoples' lifestyles take shape in the form of different practices, and the individual can be seen as a nexus of different practices (Jacobsen & Dulsrud, 2007). Thus, practices can be directly connected to each other where one practice 'spills over' into another, or one practice is the precursor or result of another (as grocery shopping practices are connected to cooking and meal time practices). Practices of food consumption are not islands but can be seen in terms of chains of practices where practices are embedded within sets of other, adjacent practices.

### 2.5.3 Concepts and indicators

SPA analyses sustainable consumption within a social practice by (1) considering the practice from a systems of provision perspective, i.e. from the ways in which practices are 'served' by particular systems of provision (from 'right to left' in the model), (2) considering the locale or physical setting of the practice, and (3) considering how the practitioner 'operates' within the practice, which meanings and actions constitute the practice ('from left to right' in the model) (Figure 2.1). In principle these three 'steps' together account for a full analysis of particular food practices like eating in a canteen, dining out or grocery shopping. The first two really constitute the 'infrastructure of sustainable food consumption', whereas the latter is meant to created insight into the practitioner.

In the next sections we will discuss these key elements.

### 2.5.4 The system of provisioning

Everyday food consumption cannot be seen separate from the infrastructures supplying products, services and the settings of consumption (canteens, snack bars, restaurants, supermarkets, etc.). SPA calls on the concept the system of provision (SoP) (Fine & Leopold, 1993) and of 'locale' (Giddens, 1984) in order to analyse such (infra)structural factors. The system of provision says something about how the food we consume is produced, processed and prepared, and how it is sourced and procured. It represents the network of actors and their social and physical infrastructures, which together make up the means through which food products and services reach the consumer within the context of a certain practice. Spaargaren *et al.* (2007a) argue that social practices can be associated with certain systems of provisioning organised in specific ways, with their own cultures of provisioning of certain types of goods and services. A provisioning system has certain rules and 'dynamics of change' which spawn different settings (locales) of consumption. The definition used in SPA is a system consisting of a certain physical infrastructure (e.g. shop floors, kitchens), providers (actors), network(s) (supply chain, business relationships), provider/company strategies, (procurement) policies and overall business/service management philosophies. In relation to sustainable consumption the subject of investigation is how sustainable food products and services are made available to consumers through the system of provision and their quality and quantity. Which specific aspects are analysed may differ depending on the social practice under consideration. In this study the following aspects have been formulated (Figure 2.1):

- the infrastructures of provisioning: supply chains, (sustainable) procurement policies, management and logistics, food preparation structures;
- the networks of provisioning: roles of and relationships between different providers (business-to-business and other) and their structuring of the system of provisioning;
- motivations, conventions and skills of providers, also in relation to sustainable development and sustainable food provisioning.

The focus lies on the workings of the system of provision that provides not only sustainable alternatives in terms of goods and services, but the narratives, images, opportunities and settings of/for sustainable food consumption behaviour. Furthermore, the system of provision (co)structures the locale; the place and time where food consumption takes place. Settings for food consumption are crucial to the kind of food experiences being made possible.

In the Netherlands most modes of provisioning in the food sector are market-based. In all sectors of the food retail and services industry we see that commercial food providers are dominant, and publicly funded food provisioning, as for instance seen in catering (e.g. work-place, education institutions, hospitals, penitentiaries) has retreated over the years. The term food provider as used in this study is meant to cover retailers, wholesalers, food service providers, processors and producers (e.g. farmers, bakers). Providers have a very distinct role in food consumption, especially today where business deals not just with the provision of products and services but with offering a range of emotion and experience of consumption as well. Providers are involved in framing products and services, and shaping the options and expectations of consumers also regarding the sustainability aspects of food (Spaargaren & Van Vliet, 2000).

Getting a grip on what role these factors play can be done by studying provider strategies in relation to sustainable provisioning. This includes how provisioning organises or orchestrates narratives and images, and opportunities and settings for engaging in sustainable food consumption. In other words, the way in which the systems of provisioning and the locale are organised to shape practices, and more specifically a sustainable practice. As Spaargaren *et al.* (2007a) note it is important to look at how providers approach consumers and consumption (practices). Adequate provisioning of sustainable provisioning is not just about the availability of sustainable food alternatives, but also about how these are being presented, how food products and services are made accessible and attractive.

### 2.5.5 The locale

The locale refers to the location and social and physical setting of a practice i.e. where, when and with whom the practice takes place. Here we find the dynamic nexus of system, infrastructure, access, use and meaning. It represents a piece of 'food culture'. Food consumption takes place in various different locales (e.g. canteens, restaurants, petrol stations, schools) which constitute important characteristics of food consumption practices. The locale tells us something about the sort of experiences which consumers have in the 'doing' of the practice. For instance, is it 'fast' or 'slow', what is the social context, i.e. with whom are people eating their meal. It reveals something about the manner in which providers and consumers are interacting. Schwartz Cohen (1987) referred to this as the consumption junction, the place where the strategies of the providing party meet the consumer (Spaargaren *et al.*, 2012). Indeed, locales can be designed and arranged in such a way as to guide consumer choice. For example, Dagevos *et al.* (2005) found that the position of organic products in supermarket shelves had an effect on consumer choices. As Giddens (1984) emphasises, locales are not 'external' to the interactions that happen at these particular segments of time-space, but instead they make possible and co-constitute practices. Exploring the locale in some detail is instrumental for understanding both the actual behaviour of participants as the strategies used by the actors within the systems of provision.

### 2.5.6 The consumer practitioner

In this study, consumers are seen as 'practitioners' who 'manage' their 'food lifestyles' within the context of daily practice(s). They apply know-how, rules-of-thumb and certain material elements. They also acquire meaning from practices, or conveying meaning onto practises (Spaargaren & Koppen, 2009), and glean a measure of (dis)satisfaction from what they do. The degree to which this is rational, strategic or calculated is left open here, and is in fact considered variable and situational in itself. A persons' 'food lifestyle' is not one of a kind or set in stone. Giddens' lifestyle concept is used here because it offers a way to conceptualise the relationship between individuals and practices. 'A lifestyle can be defined as a more or less integrated set of practices which an individual embraces, not only because such practices fulfil utilitarian needs, but because they give material form to a particular narrative of self-identity' (Giddens, 1991). In this context, consumer lifestyle is used to describe 'the way one does things', on the level of a specific social practice, or on the level of food consumption generally (i.e. involving various different practices). On the one hand this lifestyle is unique and personal. On the other hand it is to some degree always shared with others and carried out in a socio-material context, and it is variable and dynamic, i.e. dependant on situation. Individuals are not constantly busy on a conscious, discursive level to 'live their lives' and 'make and remake their lifestyles'. People are not constantly designing and planning their lifestyle in a strategic way, Giddens points out that most of the time individuals act on a level of practical consciousness (Spaargaren & Koppen, 2009).

This is the theoretical background used conceptualise the engagement of consumers in sustainable food consumption. It is reduced to three aspects which will be used in the empirical studies in this study.

#### *Consumer portfolio*

Consumer portfolio relates to the consumers as a practitioner, i.e. possessing certain knowledge and skills for 'performing' sustainable consumption (e.g. the use of organic food, eating less meat, cutting down on packaging). Consumer portfolio is the actual experience with the use of sustainable alternatives (in the context of a certain social practice), certain knowledge and skills required for the engagement in sustainable food practice (e.g. cooking skills, use of information) (Spaargaren *et al.*, 2007a).

#### *Consumer concern*

Consumer concern as formulated here relates to the perceptions on the necessity for sustainable development in food consumption and the degree to this is perceived in terms of co-responsibility (i.e. consumers see themselves as partly responsible for social and environmental effects). It also covers the more specific sustainability concerns related to food (such as concern for animal welfare or the environment) and the kinds of food alternatives perceived as offering an effective way 'counter' or 'alleviate' these concerns (e.g. eating less meat or buying organic food). This aspect can be used to understand whether there are underlying opinions and feelings of concern attached to certain practices, and the degree to which this is attached to abstract notions of consumer responsibility, specific consumer concerns or more active notions of 'doing', i.e. fining a 'solution' in terms of changing one's (buying) behaviour or practice (e.g. becoming a vegetarian because of environmental concerns).

*Practice narratives*

Practice narratives are consumers' recollections about the 'doing' of the practice', that is, recollections about daily food routines, social and physical contexts/constraints and frames of reference (norms, associations) in relation to the doing of the practice and/or engagement in sustainable alternatives (e.g. past experiences with sustainable food services). This includes people's 'stories' and evaluation of the quality and quantity of (sustainable) products and services, their experiences with various locales and the systems of provisioning. It also covers how people engage in a particular practice for example in relation to managing everyday food needs and which kinds of considerations and negotiations this includes.

### 2.5.7 Food consumption practices; change toward sustainability

Change for sustainable development requires dealing with the properties of established physical and social infrastructures of consumption, as well as the bodies which have been inscribed in certain ways by existing practices. Employing a practice-based approach means there is a potentially broader spectrum for revealing change-processes or change-pathways within the systems of provisioning, the locales of consumption, the manner in which consumers access and use artefacts, infrastructures, products and services, the social interactions and meanings within practices and the interactions between consumers and food providers. For example, changes in food practices also imply some degree of change in skills, knowledge, experience, material arrangements, images and meanings (Hand & Shove, 2004); a change in cooking practices (in order to move to a more plant-based diet for instance) would require learning new cooking skills, a learning process that many of us would not undertake on a very regular basis, certainly not in the context of one's mid-week routine.

There are various authors which have studied how practices may change over time (Hertwich, 2006; Shove & Southerton, 2000). For instance, Shove and Southerton (2000) in their study on the use of freezers by British households, showed how practices changed with the coming of the freezer. They also show how the use of the freezer has undergone three 'waves' of development and that the process through which the use of the freezer became normal involved adaptations in food practices and the scripting of practices by the freezers themselves. Spaargaren *et al.* (2007a) and others speak of the process of de-routinisation and re-routinisation (p. 54-57), where the process of change in routines is seen to occur due to crises (Hertwich, 2006; Spaargaren *et al.*, 2007a)[16] or positive triggers (used by Krom, 2008)), after which the practice normalises again, i.e. becomes

[16] Giddens speaks of fateful moments which lead to the rethinking of practices and a de-routinisation moment (Giddens, 1991).

routine. Taking consumers out of their routine thoughts and 'doings' may be part of or one of the ways in which practices can be 'opened up' for change (used by Halkier, 1999; Verbeek, 2009).[17]

Although this research does not analyse change processes empirically over some duration of time, the objective is to say something about the opportunities for and limitations of sustainable development of everyday food consumption practices. In other words, the idea is to generate insight into practices and explore where there are area's for sustainable innovations to flourish (further), or which features need to be taken into consideration for policies on sustainable innovation to work. For example, using a practice approach Verbeek (2009) found that different groups of consumers had different portfolios ('practitioners' experience/knowledge/skill) for environmentally friendly tourism mobility and made use of different transport and tourism infrastructures (p. 255). Information on environmentally friendly ways to travel are provided in such a way as to appeal to active and outdoor holidays, but fail to reach people who participate in mass tourism (p. 252). Such insights can be used by both policymakers and business.

## 2.6 Introducing the research cases and methodologies

Researching food consumption practices can be carried out using different research methodologies. In many ways, to understand the complexities of everyday food practices qualitative methodologies are most suited. Through one to one (semi-structured) in-depth interviews, situated interviews (Krom, 2008), focus groups and participant observation (Verbeek, 2009) the depths of the why's and how's of what people do can be more fully explored. They are also more suited to understand the context(s) of food practices. Since practices are being researched, it is important not just to research people, but also to research the locales where practices take place, through for example observation and situated interviews. This allows for a more thorough understanding of the social and physical infrastructure or settings of practices. Quantitative methods provide insight into patterns and trends on larger on a larger scale. It is useful when one wants to say something about certain groups of people.

In this research both quantitative and qualitative methodologies were used in order acquire two different angles on the dynamics of food practices and their participants. A survey was used to explore the engagement in sustainable food practice under Dutch consumers and a qualitative study was used for an in depth exploration of one particular food practice.

### *2.6.1 Quantitative research on consumer engagement with sustainable food practices in the Netherlands*

The survey study was carried out in order to explore the application of the SPA model to a quantitative consumer study. This survey aimed to understand Dutch consumers' engagement in sustainable food consumption by operationalising the 'left-side' of the SPA model (Section 2.5.4). This engagement was measured in terms consumers' views on 'co-responsibility', sustainability

[17] Shove contributes a slightly different view when she speaks about the birth and death of practices, where over time some practices simply disappear because certain 'ways of doing things' become obsolete and others appear.

concerns and possible solutions, as well as their everyday use of certain sustainable products (e.g. organic food) and practice (e.g. curbing food wastage). A practice perspective was brought into it by considering whether Dutch consumers with different shopping practices differ in this engagement. Many quantitative surveys on sustainable food consumption focus on the purchase of one or a few sustainable products (such as organic food products) or behaviours. In this survey a wide array of possible activities used to 'green' food practices were put before respondents. Furthermore, attention was paid to understanding the differences between respondents with different grocery shopping practices. After doing a pilot study,[18] the survey was carried out by the Contrast team and Motivaction in 2008 on a representative sample of the Dutch population (see Chapter 4 for a more detailed description of the methodology). This survey was used as the main source of data discussed in Chapter 4.

### 2.6.2 Qualitative research: the case of canteen food consumption practice

Research has paid relatively little attention to out-of-home food practices, and even less to eating in a study or work-related setting. The latter is a very specific context in everyday food consumption and various studies have shown the importance of paying attention to public food procurement (Halkier, 2010; Morgan, 1988). For these reasons it was chosen for an in depth qualitative study. Semi-structured interviews were carried out with food service providers and experts (cooks and/or catering managers working in business and public catering) and focus groups with both caterers and consumers (end-users) (for a full description of the methodology, see Chapter 4). The latter was meant to provide a comparison of the perspectives of both providers and consumers on the topic of sustainable canteen food provisioning. The focus group method allows one to delve deeper into the reasoning of respondents, their argumentations and explanations (Bloor *et al.*, 2001). In addition various canteens in the Netherlands and abroad were visited in order to be able to analyse different approaches to sustainable canteen provisioning and take into consideration the design of the locales of everyday canteen food consumption.

### 2.6.3 Literature review

Before presenting the empirical research in Chapters 4, 5 and 6, Chapter 3 provides a more in-depth discussion as to the position of the consumer in sustainable development of the agro-food sector. This chapter considers some of the most striking characteristics of food consumption today and looks at literature which supports the idea of considering sustainable food consumption via the approach of practice-theory.

---

[18] Carried out with the Contrast research project group and the NGO Stichting iNSnet (Internet Network for Sustainability) in 2007. This is a yearly monitor of sustainable consumption behaviour in the Netherlands which makes use of the database of Flycatcher Internet Research with 20,000 potential respondents.

# Chapter 3.
# Presenting the consumer in context

## 3.1 Introduction

Promoting sustainable food consumption within society is often based on the proposition that consumers are 'co-responsible' stakeholders within sustainable development (e.g. LNV, 2005; SER, 2002).[19] However, the strategies applied in both business and government policy tend to rely almost exclusively on approaches that target the individual (Shove, 2012b; Spaargaren, 2011). Often these perspectives are based on the notion that consumers are agents which need to 'take responsibility', i.e. to make responsible consumption choices, and a 'change in behaviours' is needed in order to help build a more sustainable economy. Social change is therefore framed in terms of individual behavioural change and personal responsibility (Shove, 2010). As a result, much effort is committed to the process of 'convincing' individuals without paying due attention to understanding and targeting food infrastructure and the fabric of social practices of eating and shopping for food. Furthermore, this framing results in a portrayal of consumers as socially isolated, rational actors, which, with mildly moralistic undertones, reduces sustainable consumption transition to individuals 'taking reasonability' or not. In reality, the position of the consumer and indeed of consumption in sustainable development is complex and deserves careful consideration and conceptualisation from policy and science (Jacobsen & Dulsrud, 2007). In the first place, it requires taking account of the food infrastructure or 'food landscapes'[20] in which people move, and the nature of everyday food practices in their varying contexts within society.

There are three features of everyday food consumption that have implications for the sustainable development of food consumption and the way in which the consumer is positioned in the debate on sustainable development. Foods and indeed meals are shopped for and/or eaten at various providers, in varying contexts and increasingly it seems consumers eat food (a meal) which is prepared by someone other than themselves (or someone they live with). Although these are not the only features characterising todays food consumption patterns, they are important because they provide an understanding of people's habits and consumption patterns with respect to eating and shopping for food, their daily interaction with various different locales, providers and systems of food provisioning. As we will discuss in the following sections they imply that the 'traditional' approaches towards encouraging and increasing more sustainable food practice within the everyday are no longer sufficient.

Section 3.2 discusses some features of everyday food consumption and how they are implicated in sustainable development of food consumption. Section 3.3 then touches on what these features imply for the position of the consumer. Section 3.4 considers the kinds of approaches which can

---

[19] The Dutch SER institute which advises the government on socio-economic policy has stated that sustainable consumption policy should pay attention to consumer routines, lifestyle, comfort and fashion (SER, 2002).

[20] By this I mean the food provisioning infrastructure which is made up of systems of provisioning and locales of consumption, as well as consumption patterns and practices (characteristic to certain countries, regions or areas).

be used to encourage everyday sustainable food practice, and what a practice-based approach might provide for this endeavour. The last section will show how the empirical research presented in the next chapters will deal with a number of the issues discussed in this chapter.

## 3.2 Sustainability and food consumption practice today: the consumer in context

### *3.2.1 Diversity of context and practices*

Instead of an isolated domestic-based activity, food consumption occurs in various settings and contexts or – as Schatzki would have it – sites. Especially in (sub)urban areas consumers have ample access to food services and retail. Besides the more traditional locations to eat and shop for food in city centres and office buildings, there now are café's, restaurants, fast food snack stands, food and coffee bars, vending machines and food shops in airports, train stations, schools, alongside motorways, in museums, shops, entertainment and leisure centres (Figure 3.1). Besides the 'slower' more traditional 'sitting-down' restaurants and cafés, retailers and food servicers have developed special 'food on the go' concepts. Supermarkets have opened shops in major train stations in the Netherlands, not only offering snacks and coffee but also giving travellers the opportunity to do some quick grocery shopping. Food service spending at petrol stations is growing at higher rates than spending in the restaurant sector (FSIN, 2009). These locales of 'hybrid consumption' (Dagevos & De Bakker, 2008: p. 20) mean that food consumption and provisioning are integrated in settings with various functions or practices.

The study by Cheng *et al.* (2007) provides evidence of the diversification of food consumption practices. They studied the time spent on shopping, cooking and eating in different contexts

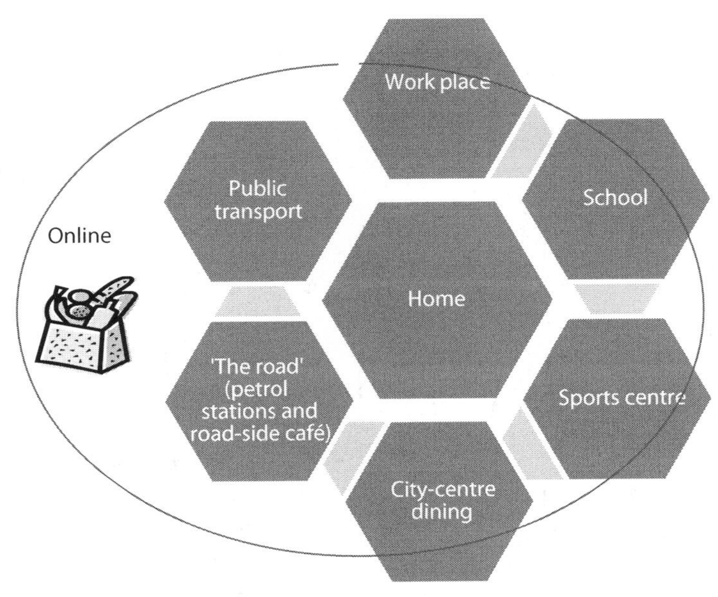


*Figure 3.1. Contexts or sites of food consumption.*

between 1975 and 2000 in the UK. During this time there was a substantial increase in the amount of time spent eating in various contexts and locations outside the home, with a greater variety in the duration of eating episodes in these contexts. In many ways the (sub-)(urban) food landscape is built in such a way as to enable the 'grabbing' of a bite to eat or some groceries throughout the day and in the context of other activities such as travelling, leisure and work. The service element within food provisioning has thus grown in importance. Food consumption can be integrated within and adapted to daily schedules and activities and in these cases eating and food shopping may no longer be the primary goal. This gives food consumption practice a flexible characteristic, less limited by tradition, etiquette, time and place. The wide spread convenient access to food has become an infrastructural property like that of transport convenience and financial convenience.

This is also apparent in the virtual access to food. The online ordering of groceries and meals has become more mainstream and some expected it to grow, possibly to 15-20% of supermarket return (ING, 2013). Besides the online services now offered by supermarket chains in the Netherlands, various businesses have appeared with different meal and grocery services, from services that provide groceries and recipes for the evening meal (HelloFresh) and others providing box-schemes with fresh organic and seasonal produce (Streekbox, Odin).

The ease with which food can be bought at 'any place and any time' has been said to result in the propagation of individualistic, flexible, informal and fragmented food consumption behaviour (Warde, 1997). Various authors suggest that the decrease in home cooking (from scratch) and 'proper' meals at the table lead to less social, less constrained and less reflexive eating captured by terms such as 'grazing' (Fine & Leopold, 1993; Strasser, 1982) and 'gastro-anomy' (Fischler, 1980). However, authors like Cheng *et al.* (2007) warn against the presumption that the commodification and individualization processes of modern consumer culture erodes social relations and 'meaningful' consumption: 'eating remains a sociable and collective practice, despite shifting temporal pressures which make the coordination of eating events within social networks more difficult' (Cheng *et al.*, 2007: p. 41). Although food consumption practices have diversified in terms of locations outside of the home, this has not greatly affect 'eating at home' practices. Time spent on cooking decreased during this period, but eating episodes within the home remained fairly stable. Some studies have shown how convenience foods are actually used to retain family meal-time, presenting a solution to the lack of time or culinary skills/interest, and despite these limiting factors being able to still realise acceptable meal quality (Carrigan *et al.*, 2006; Romani, 2005). The more traditional family meal co-exists with more flexible, individualised eating practices and convenience in food may offer a 'solution' to upholding certain traditional meal/food values.

In relation to sustainable development more pressing questions may be: Is there adequate provisioning of sustainably produced food in these different contexts? Are sustainable food alternatives available and accessible within a range of different contexts and locales of food consumption? And what is their quality? When we consider the food services sector, in the categories fast-food and 'take-away', public and business catering, as well as the restaurant and catering sector, there is a more diffuse availability of clearly identifiable and established sustainable product and service alternatives compared with the food retail sector. For instance, 13.7% of organic food sales occur within the out-of-home sector, whereas sales of organic food via supermarkets is 48.4% of the total sale of organic food (2011) (EZ, 2012). Thus access and availability play a hugely important role in determining the degree of sustainability in food practices, where there is a reliance on food provisioning outside the home. In other words, an important area for the sustainable development of food practice(s) is to be found in the varying contexts of the out-of-home food sector.

### *3.2.2 The convenience and service aspect*

Judging by the developments in food provisioning and consumption peoples' use of convenient food and shopping solutions are on the increase. 'Convenience is associated with reducing the input required from consumers in either food shopping, preparation, cooking or cleaning after the meal' (Buckley *et al.*, 2007). In home-cooking the use of ready-made food products and meals has increased by 51% since 2002 (Gfk, 2008), whilst there is some evidence that the time spent on cooking and preparing meals appears to have almost halved between 1970 and 2000 (Sluijter, 2007). Indeed whereas home food preparation has decreased in importance over the past 25 years the use of ready meals or home meal replacements and out-of-home food consumption has increased, and further growth is expected (CGE&Y, 2002; FSIN, 2008; Kamphuis *et al.*, 2006). Dutch supermarkets seem to be stocking more pre-packaged (salads, chopped vegetables and herbs, etc.), and ready-to-eat foods (sauces, meals, sandwiches, juices, etc.) than ever before. Some marketing analysts predict the blurring of the traditional retail and food services sector, with the expectation that expenditure on this kind of food services will increase over the coming years (CGE&Y, 2002; FSIN, 2008). Indeed the service aspect within food retail has become more important. Besides the range of meals on offer, food retail also provides delivery services, given a new dimension through internet shopping. Many supermarkets and other online food providers now provide online shopping services, where groceries can be selected and ordered online and either picked up at the shop or delivered at home. Internet and social media help consumers make choices, look for offers and other information. The service orientation is also seen in the fact that many online services offer recipes as well, to 'help' consumers to decide what to have for dinner, and provide cooking knowledge and skills for home-cooking.

Time scarcity, energy saving and lack of confidence in cooking skills have all been identified as reasons for choosing convenience foods (easy to prepare, ready-to-use and ready-made foods) (Carrigan *et al.*, 2006; Hollywood *et al.*, 2013; Jabs & Devine, 2006; Southerton, 2006). However, food convenience is a social construct, not necessarily just a 'set of properties of food items' (Carrigan *et al.*, 2006, p. 373). Emotional, cultural and social aspects play a role in how people define definition of food convenience. Some of the societal factors which are used to explain the increased use of convenience food are the increase in participation of women in the work force and the increase in single and two person households (Buckley, Cowan, McCarthy, & O'Sullivan, 2005). Studies have shown that the feeling of not having enough time for cooking meals is an important perceived barrier for people to realise an adequate evening meal, especially under young people, young professionals and low income mothers (IEFS, 1996; Jabs *et al.*, 2007). An Australian study on the food habits of women showed that women, who find cooking a chore, spend less than 15 minutes preparing dinner. When cooking is considered a chore, women were also more likely to eat 'fast' and/or fast food (Crawford *et al.*, 2007). In the Netherlands the aging of the population may also lead to an increased demand for convenient 'food solutions' (e.g. easy-to-prepare foods, ready-made foods) and food services. The use of convenience foods cannot only be witnessed in the domestic sphere. Many professional cooks also make use of pre-processed and pre-prepared food (FSIN, 2009). One reason for this may be that such food products makes it easier to meet strict food regulations and at the same time save on both time and labour.

The sustainable development of food practices is likely to stumble on the societal preference for convenience. Judging from the wide range of convenience foods (pre-made sauces, meals, etc.) available, many consumers choose these products above the 'cooking-from-scratch' option. Thus, furthering sustainable development within food consumption also requires the 'greening' of convenience products. Various studies show how the convenience aspect influences sustainable food consumption, how for instance convenience considerations are weighted up against considerations related to healthy eating and sustainability during grocery shopping for example (Hjelmar, 2011; Memery *et al.*, 2012). Studies show how consumers are pragmatic and look for sustainable product alternatives within their established food shopping practices (Carrington *et al.*, 2014). How they prefer shopping at a supermarket nearby and the absence of organic products here will not prompt them to change to another supermarket (Krystallis & Chryssohoidis, 2005). If sustainable alternatives are not conveniently available, in the shops and services which consumers find convenient to use, they are simply less likely to be chosen (Tanner & Kast, 2003; Weatherell *et al.*, 2003).

### 3.2.3 Less direct influence, more awareness and concern

Whereas the first half of the last century can be characterized by the increase in distance between farming and consumption, the second half can be characterized by the distancing between the consumer and food preparation (Den Hartog *et al.*, 1992)[21] With the industrialization and globalization of the food sector, food production and preparation have become a less visible part of daily experiences. For food production and farming in the Netherlands this process of detachment of food has been documented by various historical studies (Van Otterloo, 2000; Van Otterloo & De la Bruhèze, 2002; Van Otterloo & Sluyer, 2000). Schwartz-Cowan (1983) wrote how butchering, milling and canning moved out of the home to become a job done by industry and how Americans became unaccustomed to herds of cattle being driven through their city streets (p. 72). Similarly cattle markets in the Netherlands and England were moved away from town centres to peripheral areas and thus 'removed' from the everyday town life as well. Today consumers' connection with food is much more mediated through retailers, brands names and labels, and a range of 'food experts' who present the latest knowledge on food sustainability, diet-related health and culinary trends (scientists, food retailers, regulators, cooks, etc.) (Kjaernes, Harvey, *et al.*, 2007). Indeed peoples' relationship with food today makes for a complex system of knowledge based and trust based relations and perceptions.

The increase in use of food services and convenience foods imply a detachment from cooking and food handling. People have become less directly involved in the creation of their meals, and therefore in determining their taste and content. With this they are also less dependent on their

[21] In the early twentieth century a large proportion of growing, preserving, processing and preparing food took place in the domestic sphere (Den Hartog *et al.*, 1992). During the last century this relationship changed drastically due to the developments in food industrial production and processing.

own food preparation skills. As a result the end-consumer is likely to be less aware of the effort that went into making the meal and less likely to be knowledgeable about the sourcing of the meal's ingredients. The detachment between food consumption and production has been problematized in rural sociology and food sociology literature especially in relation to food sustainability and farming (Dixon & Banwell, 2004; Holloway & Kneafsey, 2004; Marsden *et al.*, 2000), but not so much in relation to cooking and food handling. However, the latter is likely to have added implications for people's relationship with food, as well their self-sufficiency in realising healthy and sustainable eating.

Although the role which convenience and food services play in providing the daily meal implies that consumers are often less directly involved in food preparation and sourcing, there is a great deal of collective and individual concern for, and interest in, food. People are increasingly aware of the problems involved in food production and the health implications of certain patterns of food consumption (e.g. consumption of processed foods containing too much sugar and salt). Thus paradoxically, whereas food production, processing and preparation has moved further away from consumers, food is maybe more than ever the subject of concern and debate (animal welfare, farmers futures, healthy eating, ecological footprint, etc.) and popular fashion ('raw food', low-gluten/low-bread diets, paleo-diets, 'kitchen on wheels', etc.). With the vast array of diversity and variety in food products and services available, as well as the vast array of topics related to food there is relatively more 'pressure' on the individual to make 'the right' choice. Van Otterloo formulates it as 'eaters are not so dependent on home/family cooks and commensality ... they can eat according to their own judgement. This situation actually puts pressure on the individual to practice more instead of less constraint' (Van Otterloo, 1990: p. 270).

## 3.3 Revaluating the 'responsibly choosing' consumer

The above issues provide ground with which to rethink the positioning of the consumer in the transition to more sustainable food systems. That is, they provide some impetus to consider the consumer in relation to the contexts of practices.

Firstly, the diversity of contexts implies that consumers eat and buy food in a range of different social-physical settings all of which have some role to play in shaping everyday food practice(s). It is therefore imperative to consider the consumer in different social practices. Different social practices of food consumption represent different traditions and 'norms' of acceptable behaviour, as well as the interaction with different systems of provisioning. This also means having different levels of access and availability of sustainable products and services. Practices and their contexts hold different 'rules of the game' ('doings and sayings') and frames of reference and meanings, and also, varying positions for the 'choosing' consumer. For instance, by now many consumers are used to finding sustainable alternatives within food retail (supermarkets and food shops), i.e. which product/labels to look for, how these products taste, which shops to visit to buy sustainable alternates. Most supermarkets carry a range of organic and Fair Trade (and other) labelled products. This is bound to be different when we look at the out-of-home food provisioning sector. In the work-place canteen consumers are dependent on what the caterer is offering. In many cases the employees have little influence on what is offered in their canteen, they have little choice but to buy what is on offer. For dining-out practices consumers are again in a different position. Here they have a larger range of choice in providers (restaurants, cafés, takeaways), although finding one that offers sustainable foods might still be difficult depending on where they are looking and may require some extra effort. In any case, looking for a sustainable alternative might not be a part of the practice-related habits of dining-out. In this case, one is more likely to be concentrating on finding a nice place to eat, or choosing food for which one is 'in the mood'.

Secondly, the position (and 'responsibility') of the consumer in sustainable development is not an isolated one; decisions on the daily meal (its sourcing, taste and contents) are often made by an array of different actors. What is often referred to, or implied as, 'individual food consumption choice' is in fact part of 'shared consumption choices'. Besides focussing on the (end) consumer it is necessary to ask who and what influences daily consumption practices. Changes in food practices do not originate from consumers only or even primarily. They result from the co-production of consumers and provisioning actors. Actors in the systems of provision have a dominant role and tend to overrule consumers in organising transitions in food practices. Providers can take a more or less pro-active strategy when it comes to including consumers in processes of change. They can provide guidance, formulate new meanings and references and last but not least can provide high or low levels of provisioning in sustainable food products and services. One should consider the roles of actors such as developers in retail and food services, buyers and food procurement managers, caterers and cooks, etc. as these actors construct sustainable practices, contexts, sustainable settings and infrastructures.

## 3.4 Changing consumption practices toward sustainability

### *3.4.1 The limited effect of information on changing practices*

Policymakers often focus on processes of 'convincing' and 'informing' consumers to make more sustainable food choices.[22] Certainly trustworthy, well-designed consumer information is essential in sustainable development as it familiarises consumers with sustainable food products and services, and informs them about things like production methods, food origin, etc. Furthermore, consumers have a fundamental right to be adequately informed and/or have adequate access to information about the contents and origins of their food. However, studies indicate that the role of information in changing practices appears limited (Bartels *et al.*, 2009; Connolly & Prothero, 2008; Halkier, 2001; Hobson, 2003). The use of information during for instance grocery food appropriation is often not part of routine behaviour and studies show that consumers often make choices according to their own experience, that is, based on personally constructed rules of thumb (INSnet, 2007). Three possible reasons for the limited effect of information on changing consumption practices are further discussed below.

Firstly, information on sustainability aspects are likely to be weighed up against other considerations and different contexts call for different considerations. Various studies have shown the inherent ambivalence involved in food consumption and the symptomatic weighing up of advantages and disadvantages, risks and benefits which come along with food choices within the

[22] For example, through intuitions like 'Het Voedingscentrum' and programmes like the taskforce for to promoting organic agriculture (Task force MBL).

context of daily lives and schedules (Connolly & Prothero, 2008; Connors *et al.*, 2001; Halkier, 1999; Macnaghten, 2003). Concerns with respect to sustainability have to be weighed up against each other (low carbon, Fair Trade), as well as against other considerations (price, availability) (Maubach *et al.*, 2009; Wilk, 2006). Halkier (1999) points out that consumption is often negotiated within a household or relationship, and that even consumers with 'green' values do not prioritise environmental consideration in all areas of practices. Furthermore, consumers may quite naturally and routinely consider the environment in some consumption practices, whereas in others they behave very differently (Dagevos, 2005; Halkier, 1999; Spaargaren *et al.*, 2007b). Non reflexive and unrefined eating moments may alternate with reflexive and well-defined eating moments, depending on the context.' Even between different spheres of activity within the daily life of one individual quality expectations may diverge considerably resulting in complex and sometimes internally contradictory, 'hybrid' consumer demands'(Renting *et al.*, 2003).

Secondly, information on sustainability aspects are more often not physically present within the context of practices of grocery shopping and eating. For instance, important information sources for consumer are NGO's (websites, flyers, etc.) (INSnet, 2007). The only information which is likely to be at hand is information on food packaging. However, Berends (2004) showed that although packaging is the number one source of food product information, not many consumers regularly read packaging information. Similarly, Maubach *et al.* (2009) found that consumers make little use of the information on packaging. Much of the information which is available to consumers, and which consumers say they use and/or trust comes from sources external to the locales of consumption practices such as from NGO's and semi-governmental organisations (e.g. Het Voedingscentrum, Milieu Centraal) via websites, television programmes and brochures.

Thirdly, the idea that consumers should and will be able to find their own information about different aspects of food suitability assumes a certain level of skill, commitment and interest. Information on food sustainability is always changing, is multi-faceted and can be contradictory. It also assumes that consumers will be able to apply it to their everyday context. Considerations about sustainability have to be integrated into existing cooking and eating routines and habits, and/or such routines actually have to be changed to become more sustainable. Often everyday shopping and cooking should only take up a limited amount of time and such contexts may not easily allow for a great deal of reflection and change. For example, studies show that consumers who are tired and/or feel hurried take in less information (Bunte *et al.*, 2008). In such a situation it is likely that reflexivity gives way to routine decision making. In one study, time constraint is the number three reason as to why consumers do not buy sustainable food products (31% of respondents) (INSnet, 2007).[23]

Changes in daily food consumption practices often require changing established habits and routines, as well as (re)negotiation with one self, partner or family member on a range of different issues (for example, convenience and price). If such changes are not 'organised' or 'instigated', then they must come from the individual him/herself. However, 'the infrastructure underlying these habits ... is often complex, negotiated and difficult to construct ... Breaking old habits and forming new ethical shopping habits require an effort beyond ethical product selection' (Carrington *et*

[23] The first reason was price considerations (72%) and the second was that providers and not consumers should concern themselves with the sustainability of products (38%).

*al.*, 2014: p. 2763). Changing consumption patterns may require disrupting (at least temporarily) existing routines, (re)considering ones habits and adapting ones views and tastes. One might have to start visiting different shops or to start learning different food preparation skills (e.g. like with vegetarian cooking). A switch to a more sustainable food routine could imply more extensive cooking and spending time finding out where to buy the sustainable ingredients. Thus, changes in consumption routines may also impact perceptions of convenience and comfort to which people are accustomed (Shove, 2003; Spaargaren *et al.*, 2007b).

### 3.4.2 Changing practices: lessons from food providers' food experience approach

It is widely recognised that food consumption has as much to do with emotion as with rational consideration and thus policymakers have indicated that the focus should be on how to 'enable' or 'seduce' consumers to make more sustainable food choices (LNV, 2009). However, instead of focusing on steering individual behaviour (or groups of individuals), encouraging sustainable food consumption could be approached through targeting practices. What becomes the focus then are aspects such as the interactions between consumers and systems of provisioning, the design and organisation of locales and the framing of food products and services (see also Chapter 2). In this way attention is paid to the consumption behaviours as rooted in socio-physical contexts.

Within the retail and food services industry this is in fact already being applied in the way that providers use 'food experience' to add value to food products and services, but also to forge a relationship of trust and familiarity with the consumer. Design and marketing are used to create a world around the product. Meanings are built into food consumption through story lines communicated via texts and images (Goffman, 1974; Mintz, 1994). Authors like Celia Lury (2004) and Gerhard Schulze ('die Erlebnisgesellshaft', 1992) point out how symbolic representations and stories of consumption have become part of the functioning of the Western market economy and society, creating a consumption culture of emotions. Both point to the importance of sensory experience above utilitarian objectives, where symbolic representations and stories become almost more important than the physical product itself (Lury, 2004; Schulze, 1992). Thus, consumption is thus not simply about the consumption of products alone, many providers aim to create an experience, a state of mind (Beardsworth & Keil, 1997).

In the food sector, and indeed in other sectors, adding meaning to consumption is achieved through the design of different 'concepts', formula's or themes whereby the kind of food offered/ served, the recipes used, the interior design of the locale, logo's, etc. are arranged in such a way that they link up to fit a conceptual whole (Hollander, 2005). A shop or restaurant formula can function like a brand, which stands for certain food qualities and the shop design communicates this formula. One might communicate low-prices, brand-less food products in a straight forward, no frills setting, another communicates vast choice of food products in a higher-price category within a comfortable shop atmosphere (Rutte & Koning, 1998). 'With shop-floor design one can add distinguishing properties. The shop-floor design is that which you communicate to the customer. And then you are talking about the emotional side of the supermarket. The functional side is not so interesting any more. Since 1985 communication determines the way shops look, that is, the technological and infrastructural aspects' (Quote from De Vries in Rutte & Koning, 1998: pp. 244-245). Both functional and symbolic aspects are intricately interwoven at the locale, complementing each other. Whether conscious or not, the locale can be thus designed to encourage certain choices and practices. Ritzer (1993) mentions how the fast food chain McDonalds has designed their restaurants in such a way as to encourage the fast through-flow of customers; making chairs uncomfortable to sit on and setting up 'drive-through's' and 'walk-through's', which all add to the general idea of 'fast' food.[24]

Such kinds of strategies are obviously also employed under food providers in the 'alternative food sector' for instance those offering sustainable food products and services (e.g. retailers/ restaurants/café's/caterers offering vegetarian and/or organic foods, locally sourced and/or seasonal food products and services). Here we find examples of how food providers come up with original ideas and concepts to frame and market sustainable food consumption. Researchers and policymakers might be able to learn from such innovative providers and see how they influence daily food consumption practices.

## 3.5 Conclusion

Although consumers should certainly not be seen as passive actors without responsibility or ability to change or affect change, this chapter has aimed to illuminate the position of the consumer by presenting arguments as to why overemphasising the role of the individual, responsible consumer in making sustainable food choices can be problematic. 'The consumer' needs to be seen in the context of today's food infrastructure and the nature of daily food consumption practices. One of the features of food practices today is that a substantial proportion of the meals that people eat is made by someone else, i.e. by the food industry, by food services or retail. This implies a substantial limitation to direct consumer influence of the daily meal. This is in contrast to the development seen in the general awareness and concern consumers have for food sustainability and health issues, animal welfare and rural welfare issues at home and abroad. The fact that the use of convenience foods and food services are so firmly embedded in everyday food practice suggests that is important that sustainable food alternatives are sufficiently available and accessible at the locations where food is ordered, shopped for and/or eaten. Food consumption takes place in varying contexts within and outside of the home, each with different situational factors influencing the outcome of food practices.

Thus, more attention should be paid to the properties of different contexts of food consumption, the organisation and design of these contexts including the food products and services provided there by providers, and used by consumers. In other words, the locales of food consumption, i.e. within the moments and settings of food consumption 'taking place', and the interactions with different actors and infrastructures within this space and time then become the topics of investigation. More focus is also needed on meal-consumption within different contexts within the food landscape (at work, in train stations, petrol stations, schools, shops, etc.).

The next chapter (Chapter 4) presents the results from a survey held amongst Dutch consumers. Here we will explore in more detail the perception consumers have of their own responsibility

[24] George Ritzer (1993) criticizes the spread of 'themed' fast foods and other franchised foods and drinks as part of the 'McDonaldisation' of society which constitutes a faceless, homogenized mode of food provisioning. See also Sharon Zukin (1996) and Alan Bryman (1999).

within sustainable development of food consumption, and their concerns and considerations in as to actively participating in more sustainable forms of consumption. Different facets of the consumers' role will be discussed; a distinction is made between the role consumers see for themselves, for consumers in general, 'in theory' and 'in practice'. This chapter will also consider how the engagement in 'sustainable food choice' differs between consumers with different everyday shopping practices. Dutch consumers are questioned on their engagement in a wide range of different sustainable food alternatives to apply in daily food practice (from combatting food wastage to eating less meat and buying organic food). They are asked to rate these on attractiveness and their availability, access and quality.

Chapter 5 and 6 takes on one of the contexts of out-of-home food consumption; the canteen lunch in a study or work-place related setting. This is a good example of ordinary daily consumption practice, in a context which has not often been considered in research. The catering sectors generally should be of huge potential importance for sustainable development since it covers a sector which influences the daily eating habits of masses of people, in different contexts (education, work, sport, travel, hospital prison, etc.). This represents a good research area to better understand the roles of food providers which are undervalued and researched, such as cooks and facility managers actors, and their role in shaping daily food practices. It presents an interesting case study to further explore how food consumption practices are 'orchestrated' and 'organised', the position of the consumer, and what opportunities lie here for transitions to more sustainable food practices.

# Chapter 4.
# Shopping for more sustainable food

## 4.1 Introduction

### *4.1.1 Towards more context in consumer studies*

Government, business and NGO's alike invest in understanding what drives food consumption and how consumers could be encouraged to make more sustainable consumer choices. In order to adapt communication/marketing messages to certain target (lifestyle) groups research is directed toward understanding costumers' values in relation to food (e.g. their taste, or how food is experienced) and their attitudes to, amongst others, sustainability (LNV, 2006), thus attempting to appeal more directly to consumers' values and consumption patterns. Much of the quantitative research on sustainable consumption is based upon attitude-behaviour perspectives (Fishbein & Ajzen, 1975) or value-based behaviour models (Burgess, 1992; de Boer *et al.*, 2007; Grunert & Juhl, 1995; Jager, 2000; Padel, 2005; Vermeir & Verbeke, 2006; Zanoli & Naspetti, 2002).[25] Here the focus lies on the link between sustainable food choice and different personal values like the Rokeach and Schwartz values (e.g. benevolence, tradition) (used in Vermeir & Verbeke, 2006, 2008) and/or food-related values (or food 'qualities' like taste, freshness, ethical considerations, etc.) (Lusk & Briggeman, 2009; Steptoe *et al.*, 1995) (used also in Bartles *et al.*, 2009 and Beekman *et al.*, 2007). Overall, many of these quantitative studies have been carried out in reference to the perceptions and/or appropriation of one or a few sustainable alternatives, notably the purchase of organic food (Dagevos, 2004a, 2004b; Grunert & Juhl, 1995; Vermeir & Verbeke, 2006; Wertheim-Heck, 2005; Zanoli & Naspetti, 2002). Many such studies give insight into consumers' generic (food) values and concerns (e.g. animal welfare, environmental care, other food quality aspects), and how these values and attitudes relate to actual consumer behaviour.

A range of different authors however argue for a more contextual orientation in consumer research (e.g. Dagevos & Munnichs, 2007; Hargreaves, 2008; SER, 2003 see also Chapter 3). This means focussing on infrastructural aspects, such as the type of food on offer in peoples' surroundings, or on food consumption in different places (e.g. at school, at work, in the supermarket, on the road), but also factors like social surroundings, food management skills, personal standards, and time constraints (Bartels *et al.*, 2009; Sijtsema *et al.*, 2002; Tanner & Kast, 2003). In marketing research, for instance, a more contextual orientation is sought through the use of the Food Related Lifestyle model (FRL). This model considers the manner in which people do their food shopping (procedural knowledge regarding the procurement of products), their cooking practices, their attitudes toward food quality aspects, food/meal-related expectations/motivations and consumption situations (the distribution of meals throughout the day and the importance of eating out) (Brunso & Grunert, 1995; Buckley *et al.*, 2005; Ryan *et al.*, 2004).

[25] For further reading the different approaches used in consumption research, see Jackson (2005) and Torjusen (2004).

### 4.1.2 Grocery shopping practices; exploring differences in engagement

Within more contextually-focussed research, various studies have indicated that where people shop for food is an important factor in sustainable food practice. Tanner & Kast (2003) found that the purchase of sustainable foods was lower for Swiss consumers with a perceived need to save time and for those consumers who shop in supermarkets. The study by Weatherell *et al.* (2003) on British consumers' interest in local foods showed that consumers' first choice for accessing local foods is via the supermarket. Such findings indicate that it is not enough to consider consumer's generic opinion/values/attitudes about societal sustainability issues and/or product alternatives. Where and how one does ones grocery shopping, and how one perceives the practice of grocery shopping is as important. As Jackson points out, shopping for food represents meaningful, socially embedded practices (Jackson *et al.*, 2006). In turn these dynamics may not only determine people's access to 'sustainable' food alternatives, but also their meanings, frames of reference and expectations in relation to food/food provisioning quality and sustainable development within the food domain in general.

This chapter discusses the results from a large scale, exploratory survey which aims in part to take shopping-practice into account. Numerous consumer surveys result in constructing lifestyle groups and typologies of consumers on the basis of various life values, food-related values and behaviours (e.g. Beekman *et al.*, 2007), but not many consider actual food practice(s). Besides investigating Dutch consumers' general engagement in sustainable food consumption, this survey compares this engagement between respondents with different shopping practices. Furthermore, whereas studies on food consumption often consider sustainable consumption in relation to a limited range of 'sustainable food behaviours', defined in various ways (e.g. organic food and meat consumption), this survey includes a wide range of sustainable food alternatives both product alternatives and practical alternatives. In this way it employs a broader definition of 'sustainable food behaviours', in this case referred to as the engagement in sustainable consumption.

The research questions posed were:

- How do Dutch consumers view their role in sustainable development of food consumption and to what extent do they engage in a range of different sustainable food alternatives?
- How do consumers with different shopping practices differ in their inclination toward and engagement in sustainable food consumption?
- What can we learn from the use of the SPA framework for a quantitative survey as carried out in this study?

First we will discuss the survey methodology (Section 4.2) and set-up (Section 4.3). Then Sections 4.4 through to 4.8 present the results of the survey. Section 4.9 provides a discussion and Section 4.10 the concluding remarks.

## 4.2 Methodology

### 4.2.1 Use of the social practice approach in the survey set-up

As discussed in the theoretical chapter (Chapter 2) the analysis of consumption is approached from Giddens' lifestyle concept. Here meanings and references (opinions, concerns), experiences and knowledge, define the 'practitioner' and are seen as a way to characterise consumer engagement in sustainable food consumption (Figure 4.1).

In the survey these aspects were measured primarily in relation to the food domain instead of placing them under headings of specific social practices, i.e. explicitly placing questions within contexts such as eating out, grocery shopping, etc.[26] However, questions were posed on where respondents do their grocery shopping in order to then investigate possible differences in engagement within the food domain. Also, respondents' use of various sustainable alternatives (portfolio) was measured in relation to three different contexts (grocery shopping, the canteen and dining out).

### 4.2.2 Co-responsibility

Consumer co-responsibility measures the extent to which respondents perceive consumers and themselves as partly responsible for the environmental impacts of food choices and respondents' own active participation in making sustainable food choices. The manner in which consumers view their responsibility may vary between finding sustainability issues important, and actually

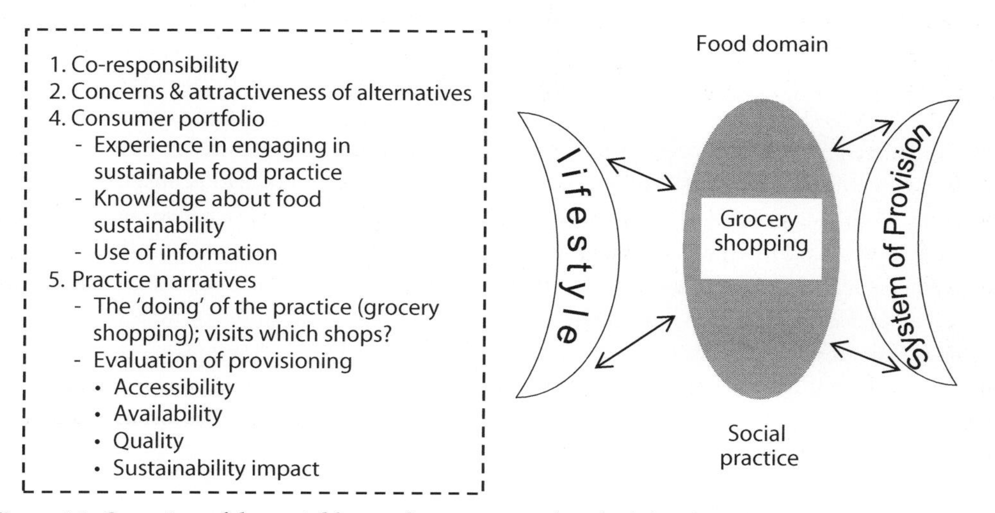


*Figure 4.1. Overview of the variables used to operationalise the left side of the social practices model (adapted from Spaargaren* et al., *2007a).*

[26] With the original intention of being able to compare different domains with each other (so, mobility, food consumption, touristic travelling and housing). Not further discussed in this study.

engaging actively by for instance buying sustainable products. In the survey the co-responsibility questions were set up in such a way as to delve deeper into the these different facets.

### 4.2.3 Concerns and the attractiveness of food alternatives

Various studies have been carried out to uncover the motives behind food choice. These may be expressed in terms of certain use-related aspects (convenience) or qualities such as price, taste and other 'food-values' (Beekman *et al.*, 2007; Brunso & Grunert, 1995; Lockie *et al.*, 2002; Makatouni, 2002; Steptoe *et al.*, 1995; Torjusen, 2004). Much research has been dedicated to understanding the motivations behind the purchase of organic food (Lockie *et al.*, 2002; Makatouni, 2002; Torjusen, 2004). On the subject of food related values or qualities many studies reveal that perception of price,[27] safety, taste, healthiness, appeal and convenience are considered the most important factors that influence food choice. Research on organic food consumption consistently shows similar outcomes, namely that health, taste and price are considered the most important concerns (Baltussen, 2006; Bartels *et al.*, 2009; Koens, 2006; Lockie *et al.*, 2002; Shepherd *et al.*, 2005). This suggests the universal importance of these aspects and thus food safety, price, health concerns and taste(s) omitted from the survey. It was assumed that such factors would always be at the top of the list of concerns.

Equally important to understanding concerns of consumers is finding out what ways of 'greening' food practices they consider attractive. There is not much known about what kinds of food alternatives Dutch consumers find attractive in cases where respondents can choose from a wide variety of clearly defined and recognisable alternatives. Also, it was investigated which concerns respondents associate with different sustainable alternatives, thus making a connection between food-related sustainability concerns and sustainable alternatives.

### 4.2.4 Consumer portfolio; knowledge and experience

Sustainable consumption portfolio refers to people's accumulated knowledge and experience with sustainable provisioning (food products and services).[28] The concept of consumer portfolio is based on the idea of seeing consumers as practitioners (Warde, 2005); consumers act not as isolated individuals, but practice consumption in a social and physical context. In these contexts consumers employ certain know-how in order to function and participate within that particular practice. The basic assumption is that know-how and experience are acquired from, and/or required to, engage in more sustainable forms of food consumption. This know-how and experience base is likely to differ between practices (something which is not investigated here, as mentioned on the previous page). The following aspects were included to measure the sustainable consumer portfolio.

[27] It appears that consumers often mention price as one of the most important factors determining their food choice, but in reality many factors play a role in people's price perceptions (e.g. Dagevos *et al.*, 2005).

[28] Or, the competence and familiarity with a product, service, activity or artefact within the context of a social practice (Spaargaren *et al.*, 2007b).

Firstly, peoples' experience in engaging in sustainable food practice, which was based on the frequency of use of certain food product alternatives (e.g. organic food) but also on the engagement of certain 'practical alternatives' like eating less meat and eating seasonal foods (Table 4.1).

Secondly, consumers' knowledge about food sustainability issues and how to make daily food practices more sustainable. Tanner & Kast (2003) take a similar approach. They define food knowledge as factual knowledge and action related knowledge (Tanner & Kast, 2003 citing Schahn & Holzer (1990)). Action-related knowledge refers to information about possible actions which was measured in relation to the respondent's idea about what kinds of actions/products are more harmful to the environment than others. Factual knowledge relates to 'knowledge about definitions and causes/consequences of environmental problems' (Tanner & Kast, 2003: p. 886).

Thirdly, the use of information, that is, to what extent consumers consult different sources of information.

*Table 4.1. Overview of the sustainable alternatives under investigation within the survey.*

| | **Sustainable alternatives** |
|---|---|
| Practical alternatives | Cutting down on food wastage |
| | Cutting down on packaging (relates to fresh produce, i.e. fruits and vegetables) |
| | Eating less meat (max. of three times a week) or no meat (vegetarian diet) |
| | Eating seasonal foods (relates to fresh produce, i.e. fruits and vegetables)[1] |
| | Eating locally/regionally produced foods (relates to fresh produce, i.e. fruits and vegetables) |
| Product alternatives (most are certified, with label) | Buying foods with compostable packaging |
| | Eating organic food |
| | Eating meat with a label for animal and/or environmental production |
| | Eating sustainable farmed/caught fish |
| | Eating Fair Trade products |
| | Buying food products with a Dutch Eco-label (Milieukeur) |

[1] One could argue that seasonal foods are also product alternatives. However, it was assumed that in general they do not have specific labels, like for instance local and regional produce may have. Also, seasonal produce can be recognised without a specific label.

### 4.2.5 Practice narrative: 'doings' and 'sayings' of the practice

Practice narrative refers to peoples' personal accounts of practices ('doings'; and 'sayings'), that is, where, when, how and with whom practices are carried out. These can include a variety of different aspects: experiences, frames of references, quality expectations, norms and standards related to food. In this survey the practice narrative centres on:

1. Consumers' evaluation of the provisioning of the various sustainable food alternatives (Table 4.1) (within the food domain, not practice specific).
2. Grocery shopping practices; where consumers shop for food.

The evaluation of sustainable food provisioning can potentially cover a large range of aspects, but has been narrowed down to the *accessibility*, *availability*, *quality* and *sustainability impact* of sustainable food alternatives. Accessibility refers to access to sustainable alternatives, or products which are needed to engage in the alternative, or, the ease of use/engagement in that particular alternative. Availability refers to the presence/absence of products, or the range of products available. Quality refers to product quality, and/or the user friendliness and trustworthy of the products or engagements. The contribution to sustainability refers to the extent to which respondents feel that the alternative represents a good way to engage in sustainable food practice.

Where people shop for food says something about the kinds of locales and systems of provisioning they come into contact with, what kinds of food products they will have access to, how these foods are presented, priced, etc. and what kinds of sustainable food alternatives are available. Shopping in different (sets of) shops might therefore be classed as different social practices, or sub-practices. It was investigated whether respondents with different grocery shopping practices also differ in their co-responsibility, perceptions on the attractiveness of alternatives, portfolio and evaluation of alternatives.

## 4.3 The survey

### *4.3.1 Survey set-up*

The survey was carried out amongst Dutch consumers (n=2,288) in July and August 2008 and is the result of the cooperation between the Contrast Project team and Motivaction, a commercial market research organisation which specialises in researching consumer behaviour. The Contrast project team (i.e. consisting of the Environmental Policy Group at Wageningen University, the Dutch Environmental Assessment Agency (PBL) and the Agricultural Economics Institute (LEI)) constructed the survey questionnaire and Motivaction provided access to its pool of respondents. Motivaction has its own on-line panel called StemPunt, which consist of about 100,000 members. StemPunt gives representative samples of the Dutch population between the age of 18 and 65 years of age based technique of propensity-sampling, where respondents are selected not only on the basis of socio-demographic characteristics, but also on the basis of relevant lifestyle characteristics.

The survey on food consumption was conducted alongside the surveys on four other consumption domains being investigated in the Contrast project: mobility, touristic mobility, housing and clothing (see Appendix I for a copy of the questionnaire). The on-line questionnaire was divided into three parts in order to minimise fall out:

- Part 1: questionnaire on general environmental issues and everyday mobility (n=2,242).
- Part 2: questionnaire clothing and tourism (n=2,302).
- Part 3: questionnaire on housing and food (n=2,288).

Respondents received the questionnaires at intervals, receiving an invitation per e-mail for each of the three questionnaires. All respondents received all three questionnaires with fall out differing per questionnaire. The total sample size was 2,906 respondents.

Each survey began with an explanation of the term sustainability as the term is often confusing for many Dutch speakers (Bakker, 2007; Winter *et al.*, 2008).[29] Overall, care was taken as to use uncomplicated language. Questions consisted of 4-6 point Likert scales. Table 4.2 provides an overview of the survey questions. Within the SPSS database, Likert scales were coded into values, where for example totally disagree/totally unattractive was given value 1 and totally agree/very attractive was given value 5. Where required Likert scales were recoded into appropriate values.

Where appropriate results are mentioned from the pilot survey carried out by the Contrast research project group and the NGO Stichting iNSnet (Internet Network for Sustainability) in 2007. This survey functioned as a pilot study for the Contrast project. The iNSnet survey is a yearly monitor of sustainable consumption behaviour in the Netherlands which makes use the database of Flycatcher Internet Research with 20,000 potential respondents (iNSnet, 2007).

### 4.3.2 Questions and routing

For co-responsibility respondents were asked how they perceive their role and the role of consumers generally, with respect to sustainable food consumption. Then respondents were asked to rate the importance of various sustainability issues[30] and following this, to rate the attractiveness of the various sustainable food alternatives (as shown in Table 4.1). For the three most highly rated alternatives respondents were asked to give reasons as to why they found these options attractive (using routing). Respondents were asked to choose answers from a list of concerns (as many answers as they wanted) (Appendix 1). This list included one open answer, so that respondents could fill in a reason/concern of their own choice and relevant concerns for each specific alternative were included.

With respect to consumer portfolio, whereas all respondents received all the questions on food knowledge and use of information, each respondent received only two questions on the frequency with which they make use of two of the food alternatives (given randomly). For food knowledge, a number of general food questions were posed on the sustainability implications of the agro-food and fishing industry. Two more practice orientated questions tested respondents' knowledge on being able to distinguish food products which are more sustainable and how meals can be made more sustainable. The use of information measured how frequently a certain source is used for the purpose of making a more sustainable food choice.

For practice narratives, respondents received questions on their grocery shopping practices. The questions were directed at how frequently respondents shop for food at: the same supermarket, different supermarkets, natural food shops or organic food shops, farmers markets, local shops (i.e. bakers, butchers and green grocers), shops where there are special offers and websites on the

[29] In Dutch the word 'sustainable' also means 'durable'.

[30] The consideration of concerns such as health, food safety, taste and price were left out as explained in Section 4.2.1.

*Table 4.2. Overview of the survey (in order of how respondents received the questions).*

| Likert scale | Questions | |
|---|---|---|
| Co-responsibility (totally disagree, disagree, neutral, agree, totally agree, don't know) | To what extent do you agree/ disagree with the following: | • consumers are partly responsible for the environmental problems caused by the food products they buy;<br>• I think it is important that food products are sustainable;<br>• consumers should actively choose for sustainable alternatives;<br>• I make sure the food I buy is sustainable;<br>• conventional food products are sustainable enough;<br>• care for the environment has nothing to do with the food products I buy. |
| Attractiveness of alternatives (very unattractive, unattractive, neutral, attractive very attractive, don't know/no opinion) + reasons why attractive with routing | How attractive do you find ... | buying organic food, etc. |
| | Why do you find it attractive? | differs for each alternative, see Appendix 1 |
| Shopping practice (does not apply to me, neutral, does apply to me. don't know/no opinion) | In my case/in our case shopping is done at ... | • mostly the same supermarket;<br>• various different supermarkets;<br>• shop other than the supermarket: ...;<br>• shop/ supermarket where there are special offers;<br>• small independent shops (butcher, baker, green grocer);<br>• natural food shops/organic shops/supermarkets;<br>• farmers market;<br>• internet. |
| Food knowledge (know nothing about it, know little about it, know something about it, know a lot about it) | How much do you know about ... | • how to make meals more sustainable by cooking with different ingredients;<br>• which food products are sustainable and which are not;<br>• the problems agriculture and food trade cause in third world countries;<br>• the environmental problems caused by the meat and the food industry;<br>• the environmental problems caused by agriculture;<br>• the extinction of fish because of large scale fishing. |

| | | |
|---|---|---|
| Experience (frequency of use/ purchase) (never but recognize it/know about it, never & never heard of it, now and then, regularly, almost always) + routing, questions on 2 alternatives | How often do you ... | buy organic food, etc. (per sustainable alternative) NB: Packaging question includes: cutting down on food wastage and use of compostable packaging |
| | In which situation do you make sure that you do this ... | • during grocery shopping;<br>• in the canteen at work;<br>• whilst eating out in restaurants/cafes;<br>• none of these. |
| Use of information (never, now and then, often, always) | I get information about food sustainability from ... | • shop staff;<br>• information present on the shop floor;<br>• NGO's;<br>• consumer organisations;<br>• television programmes;<br>• labels on packaging;<br>• family/friends;<br>• I go on my own experience and knowledge. |
| Evaluation of provisioning (totally disagree, disagree, neutral, agree, totally agree, don't know) + routing: questions on 2 alternatives (same as with experience) | To what extent do you agree/ disagree with the following ... | • the alternative is easily accessible/is easy to use;<br>• there is enough choice (range of products);<br>• the quality is good and/or it is user friendly and/ or it is trustworthy;<br>• the alternative is a good way to contribute to sustainability.<br>NB: Packaging question includes: cutting down on packaging and use of compostable packaging |

internet. However, for the logic of the survey, these questions were asked earlier on (before the questions on portfolio).

The second set of practice narrative questions were to ask respondents to evaluate the provisioning of different sustainable food alternatives based on their experience with these alternatives. In the survey respondents were asked to evaluate the same alternatives as they had received portfolio questions about. For instance, if a respondent received portfolio questions on organic food and eating less meat for instance, he/she would then also be asked to evaluate these alternatives on accessibility, availability, quality, and contribution to sustainability. Thus, each respondent answered questions on a different set of alternatives (Figure 4.2).

## 4.4 Analysis of results

### *4.4.1 Descriptive statistics and grouping of respondents on shopping practice*

Through the use of descriptive statistics each of the variables was analysed individually. The figures indicate frequency scores on Likert-scales (Table 4.2 and Appendix 1). Respondents

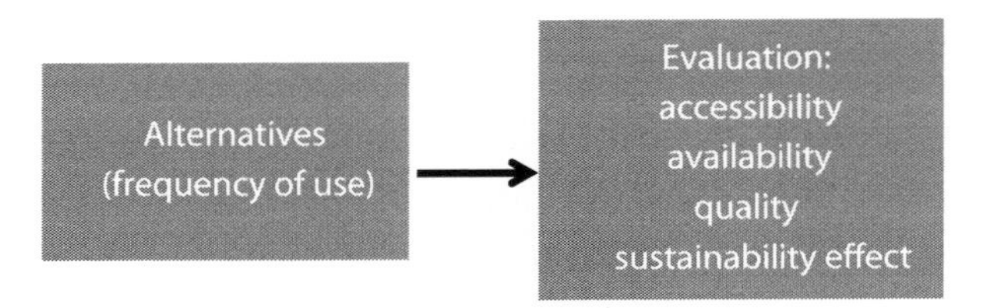


*Figure 4.2. Evaluation of alternatives on the basis of accessibility, availability, quality and sustainability effect.*

received questions on their grocery shopping practices in order to investigate whether consumers with different shopping practices differ with respect to perceived co-responsibility, perceived attractiveness of sustainable food alternatives, experience with such alternatives and knowledge (portfolio), and their evaluation of provisioning of sustainable food alternatives (Figure 4.3).

In order to do this the data on shopping channels was used to group respondents into seven different groups:

1. Those respondents who only shop at one and the same or different supermarkets (n=846);
2. Those who only shop at alternative shops, i.e. natural food shops, organic food shops, farmers markets and farm shops (the alternative channel) (n=48);
3. Those who only shop at conventional specialised shops, like butchers, bakers and green grocers (n=172);
4. Those who shop at the supermarket and alternative shops (n=106);
5. Those who shop those who shop at the supermarket and conventional specialised shops (n=434);
6. Those who shop at alternative shops and conventional specialised shops (n=71);
7. Those who shop at all three types of shops (n=143).

As the number of respondents who shop exclusively in alterative shops was relatively small (n=48), analyses were also carried out using six instead of seven groups. To make six groups respondents of group 2 and 6 were pooled together. Thus the alternative shopping practice group came to include those respondents who also shop at conventional specialists (baker, butcher and green grocer). For some analysis this gave better results (larger differences in the post hoc SNK test). This was the case for perceived attractiveness and evaluation of provisioning for the post hoc tests (SNK).

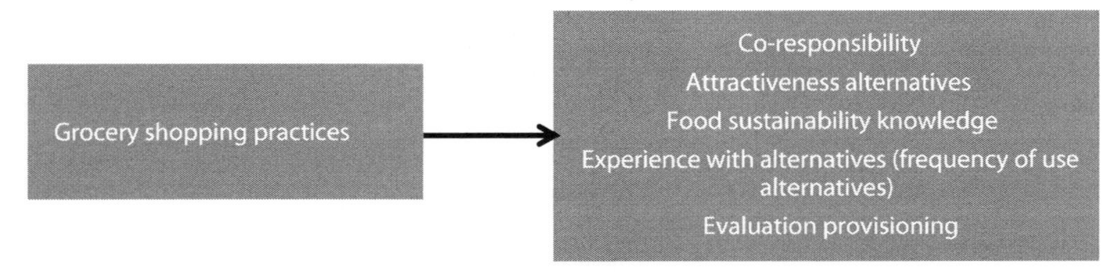


*Figure 4.3. Overview of factors included for analyses of respondents with different shopping practices.*

### 4.4.2 Statistical tests

Co-responsibility and food sustainability knowledge are constructs and in order to measure their reliability Cronbach's alpha (CA) was used. This value measures the internal consistence of a scale and thus tests if the statements used in the questionnaire measure the same underlying construct, in this case food knowledge and opinions on co-responsibility. Reliability is assumed at a CA coefficient of 0.70 (Pallant, 2005). In order to come to one overall value measuring co-responsibility and one measuring food knowledge, scores on questions related to these subjects were totalled. The statements used to measure co-responsibility (Table 4.1) were re-coded to appropriate values (i.e. making sure totally disagree = 1 to totally agree = 5).

Differences between shopping practices were tested for using ANOVA one-way comparison of means, with port hoc tests (SNK). Additional general linear (univariate) models and SNK post hoc tests were used to produce effect sizes ($Eta^2$) in order to judge the extent to which shopping practices (groups) differed (0.01=small effect, 0.06=moderate effect and 0.14=large effect) (Pallant, 2005). Statistical analyses were carried out using IBM SPSS Statistics 21.

## 4.5 Shopping practices; where people shop

The results show that around 60% of respondents visit the same supermarket for their daily groceries.[31] Almost the same number of respondents indicate that they do their grocery shopping at different supermarkets (Figure 4.4) and visit shops/supermarkets where food products are on offer. A much smaller percentage of respondents visit natural food/organic shops (9%) and farmers/ farmer shops (17.5%). The fact that so many consumers shop where food products are on offer reveals some of the thought and effort which can go into grocery shopping. Another consumer

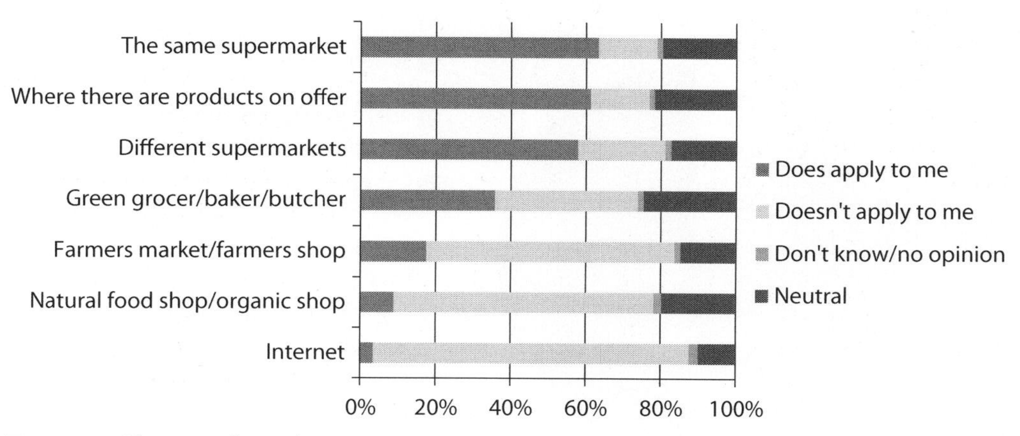


*Figure 4.4. Shopping channels.*

[31] In comparison, LEI (2006) reports that in the Netherlands 80% of food is bought at supermarkets.

survey held in 2008 found that almost a third of Dutch consumers make regular use of supermarket flyers to look for specials (survey by Q&A Research & Consultancy) (Distrifood, 2008).

After dividing respondents into various groups we see that around 46% of respondents shop exclusively in supermarkets, that is, at one or more supermarket chains (Figure 4.5). Thus, apparently almost half of Dutch consumers are solely depend on supermarkets for their grocery shopping. In comparison, only 3% shop exclusively in alternative shops such as organic and natural food shops, and farmers shops and markets. Of all respondents 9% shop exclusively at conventional specialists like butchers, bakers and green grocers. The remaining respondents (42%) shop at different combinations of the three shopping channels (i.e. supermarket, alternative shops and conventional specialists).

## 4.6 Consumer co-responsibility

Judging from respondents' disagreement of the statement that food consumption has nothing to do with the environment (70% disagreed or totally disagreed, see Figure 4.6), a large proportion of Dutch consumers are aware of the link between environmental degradation and food consumption. 55% of the respondents agree or strongly agree that consumers are responsible for the environmental burden that food products cause and consider it important that food is sustainable. Respondents seem less convinced about their own active role and whether or not conventional food products are already sustainable enough. Only 23-35% of the respondents agree or strongly agree with the statements on respondents' own active involvement in making food consumption more sustainable (statement: 'I make sure that the food I buy is sustainable'), and 20% of respondents agree or totally agree with the statement 'conventional food products are sustainable enough'.

Remarkably, the results also show a relatively high degree of neutral answers (e.g. average 36% of respondents answered 'neutral'), especially for 'consumers should actively choose sustainable food alternatives' and 'I make sure the food I buy is sustainable' (>40%). Thus, overall it seems that

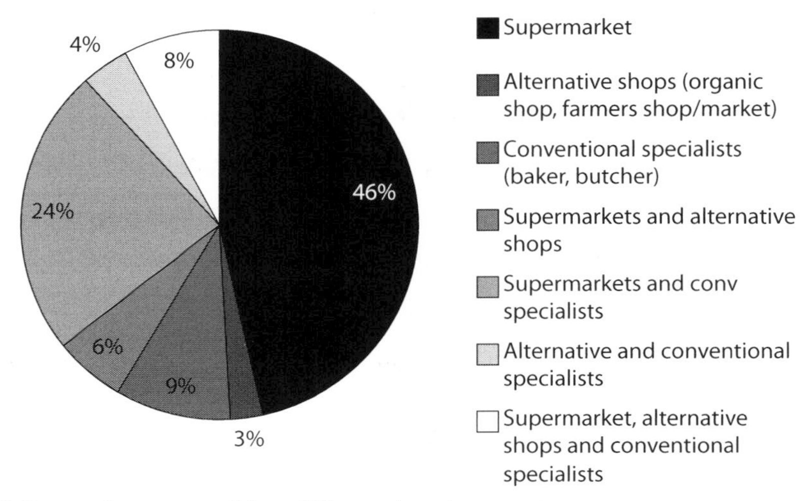


*Figure 4.5. Respondents grouped into different shopping practice types.*

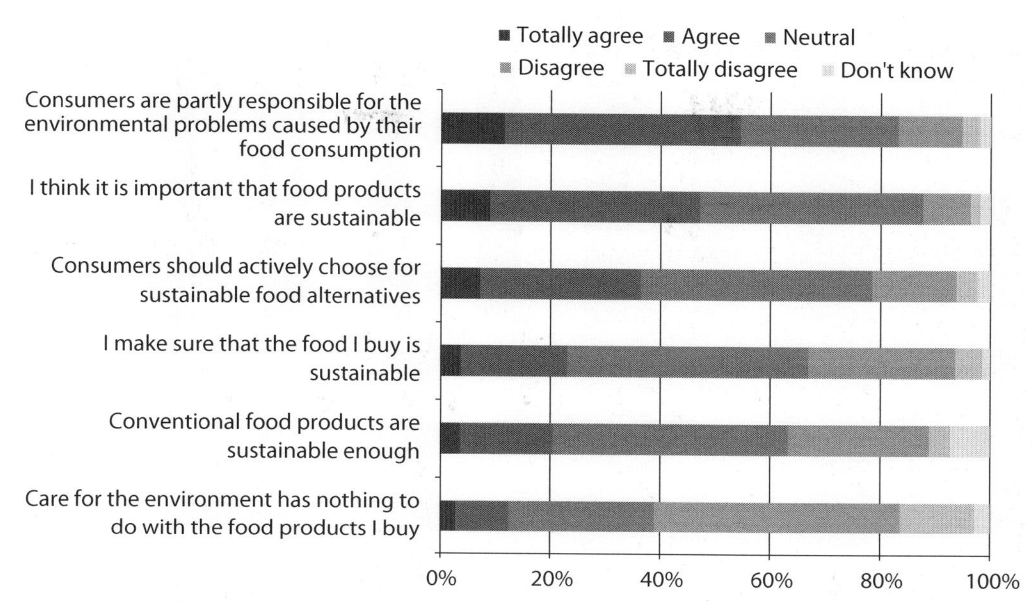


*Figure 4.6. Respondents' views on the co-responsibility of consumers in relation to sustainable food consumption.*

respondents are less opinionated/less decided about their own behaviour compared to the more general statements on the role of consumers and consumption (30% chose neutral for 'Consumers are partly responsible ...' and 'Care for the environment has nothing to do with ...'). In addition, over half of respondents answered either 'neutral' or 'don't know' on whether conventional food products are sustainable enough. This again seems indicative of a large degree of ambiguity and/or a lack of sense of urgency surrounding a sustainable food choice.

The total score on co-responsibility is meant to give an indication as to whether consumers perceive themselves as responsible and active agents within sustainable development versus as more passive bystanders (Figure 4.7). Statements used in the questionnaire were relatively reliable indicators for the measure of co-responsibility (Cronbach's alpha of 0.7 in Table 4.3b). Overall the results show that respondents with different shopping practices differ in co-responsibility but maybe not as markedly as one might expect (effect size $Eta^2 = 0.05$, small-moderate). Although respondents who visit alternative shops (organic/natural food shops, farmers shops/markets, exclusively or in combination with the other shops) score significantly higher on co-responsibility than respondents who do not shop at alterative shops, post hoc analysis shows that this difference is only small to moderate (Table 4.3).

When we look at the scores on the individual statements we see that there is no significant difference in how respondents with different shopping practices respond to the statement 'The environment has nothing to do with the environment' (Table 4.4). The most difference is seen on the statement 'I make sure the food I buy is sustainable'. Respondents who shop exclusively in supermarkets are less inclined to make sure the food they buy is sustainable compared with respondents with other kinds of shopping practices ($Eta^2=0.083$, effect size is moderate-large).

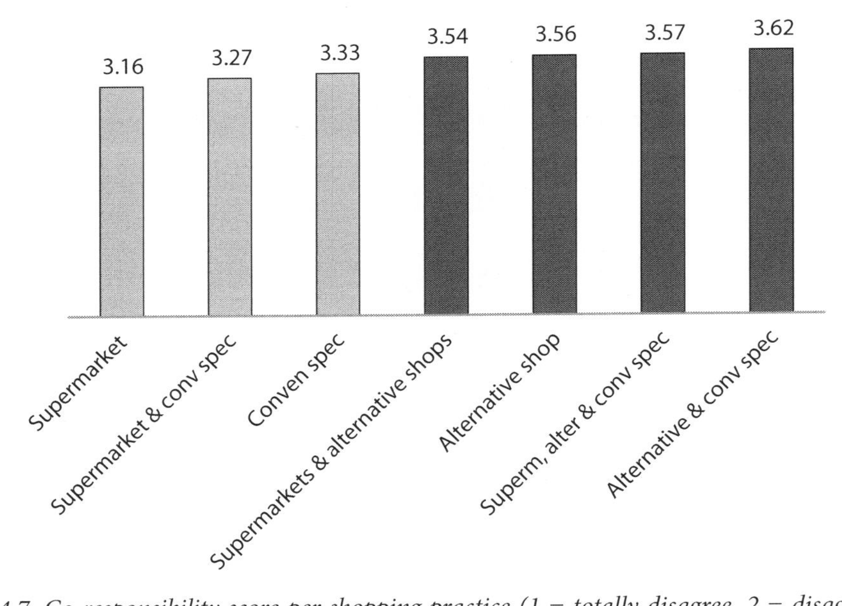


*Figure 4.7. Co-responsibility score per shopping practice (1 = totally disagree, 2 = disagree, 3 = neutral,4 = agree, 5 = totally agree).*

*Table 4.3. Respondents mean scores on co-responsibility per shopping type (7 groups).*

| **Co-responsibility** | **n** | **Mean**[1] | **Standard deviation** |
|---|---|---|---|
| Supermarket | 810 | 3.16 | 0.65 |
| Alternative | 48 | **3.56** | 0.79 |
| Conventional specialist | 163 | 3.32 | 0.65 |
| Supermarket and alternative | 102 | **3.54** | 0.72 |
| Supermarket and conventional specialist | 418 | 3.27 | 0.66 |
| Alternative and conventional specialist | 68 | **3.62** | 0.60 |
| All three | 138 | 3.57 | 0.66 |
| **Cronbach's alpha** | **Eta²** | **ANOVA** | |
| | | **F** | **Sig.** |
| 0.70 | 0.05 (small-moderate) | 16.25 | 0.00 |

[1] ANOVA, SNK post hoc: significantly higher scores shown in bold (Sig. 0.00).

*Table 4.4. Mean scores on different co-responsibility statements showing significant differences between respondents who shop exclusively at the supermarket and respondents with other shopping practices.* [1]

| | Conventional food products are already sustainable enough. | The environment has nothing to do with the food I buy in the shop | I find it important that food products are sustainable | I make sure the food I buy is sustainable | Consumers are partly responsible for the environmental problems that are caused by the food products they buy | Consumers should actively choose sustainable alternatives (like organic, Fair Trade products, etc.) | n |
|---|---|---|---|---|---|---|---|
| Supermarket | 2.78 | 3.50 | 3.36 | 2.66 | 3.44 | 3.09 | 846 |
| Alternative | 3.29 | 3.71 | 3.71 | 3.27 | 3.71 | 3.67 | 48 |
| Conv. spec. | 2.95 | 3.46 | 3.55 | 3.12 | 3.50 | 3.30 | 172 |
| Supermarket & alternative | 3.30 | 3.65 | 3.74 | 3.24 | 3.66 | 3.48 | 106 |
| Supermarket & conv. spec. | 2.81 | 3.51 | 3.48 | 2.98 | 3.55 | 3.17 | 434 |
| Alternative & conv. spec. | 3.32 | 3.62 | 3.84 | 3.34 | 3.78 | 3.66 | 71 |
| All three | 3.12 | 3.61 | 3.80 | 3.42 | 3.78 | 3.62 | 143 |
| F | 7.44 | 0.87 | 10.72 | 27.07 | 4.25 | 12.72 | |
| Sig. | 0.00 | 0.52 | 0.00 | 0.00 | 0.00 | 0.00 | |
| $Eta^2$ | 0.003-0.041 (small-moderate) | 0.003-0.041 (small-moderate) | 0.003-0.041 (small-moderate) | 0.083 (moderate-large) | 0.003-0.041 (small-moderate) | 0.003-0.041 (small-moderate) | |

[1] ANOVA, SNK post hoc: significantly higher scores shown in bold (Sig. 0.00).

For all other statements the differences are significant but small ($Eta^2$=0.003-0.041). Overall, the least differences are found for the general statements on the link between consumption and the environment and the responsibility of consumers ('consumers are partly responsible for ...'). Statements on active choice for sustainable alternatives and making sure the food one buys is sustainable show more variation between respondents with different shopping practices (see also Figure 4.8).

## 4.7 Sustainability concerns and attractiveness of food alternatives

### *4.7.1 Concerns and attractiveness*

Amongst respondents the three most important sustainability issues are Fair Trade (86%), animal friendliness (82%) and environmentally friendliness (81%) (Figure 4.9). These results are similar to other survey findings where social and human welfare issues are at the top concerns (INSnet, 2007; Koens, 2006). The fact that animal welfare is considered important by so many respondents

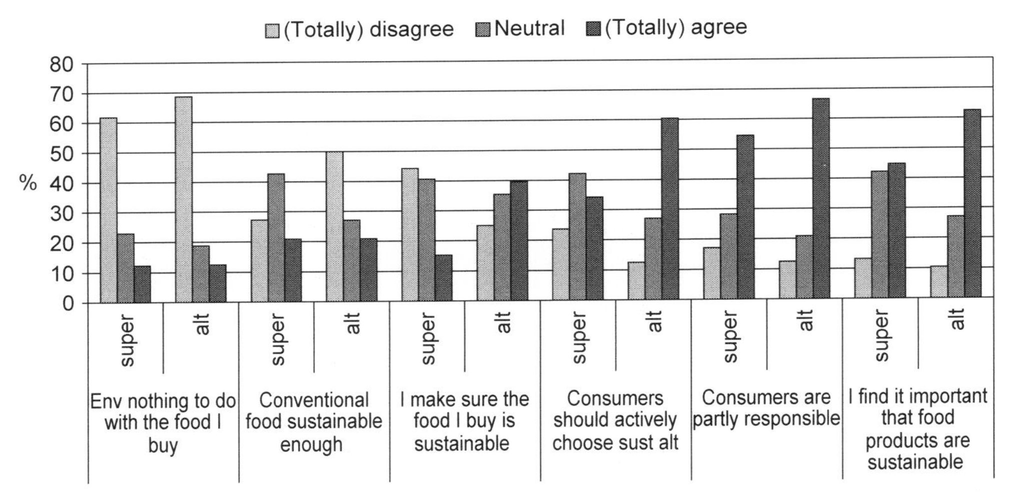


*Figure 4.8. Results on co-responsibility statements for respondents who shop exclusively at supermarkets and respondents who shop exclusively at alternative shops.*

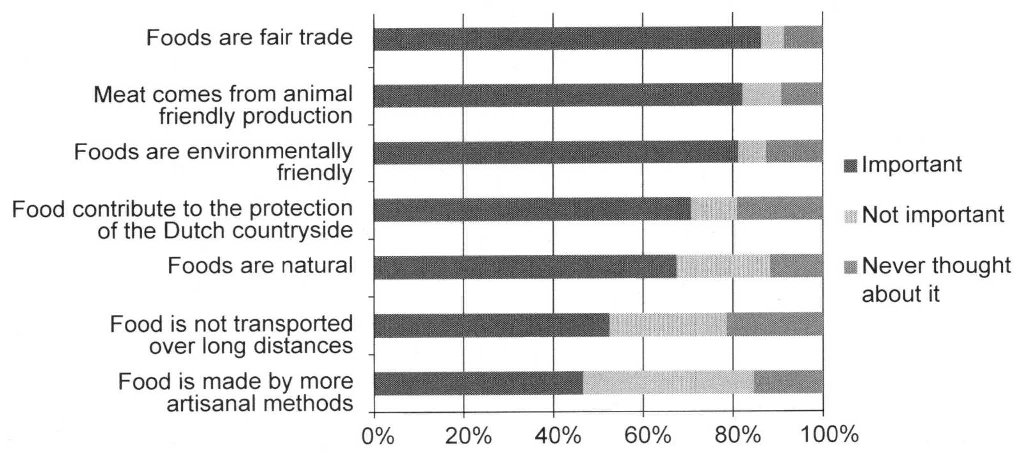


*Figure 4.9. Importance of different food sustainability concerns (e.g. 'Foods are natural' was explained as foods are not sprayed with pesticides or do not contain artificial additives).*

is not surprising in view of the attention on animal welfare. In 2006 a party for animal welfare ('Party for the Animals') was voted into the Dutch parliament and in the Netherlands a number of Dutch NGO's have long-running campaigns on animal welfare and the sustainable reform of the Dutch animal farming sector. 'Eurobarometer' surveys, e.g. European Commission (EC, 2005) and studies from the EU-funded Welfare Quality project (Kjaernes, Roe, *et al.*, 2007) show that European citizens show a strong commitment to animal welfare.

This said, eating less/no meat is considered the least attractive sustainable option, with almost 40% of respondents finding it unattractive or very unattractive (Figure 4.10). This is a relatively high percentage since the number of respondents finding the other alternatives unattractive and very unattractive remained under 20%. Eating less or no meat is considerably less attractive than buying meat with a special label (relating to animal friendly production and/or environmentally friendly production). Using less packaging, wasting less food and buying seasonal fruits and vegetables are considered the most attractive sustainable food alternatives (Figure 4.10). Buying organic food and eating less or no meat (vegetarian diet) are considered the least attractive.

### 4.7.2 Perceived attractiveness and shopping practice

Whether having a different shopping practice meant that respondents rated the attractiveness of alternatives differently was measured in two ways. Firstly by using a total score of attractiveness and secondly by looking at how respondents with different shopping practices rated each sustainable food alternative separately. The total attractiveness score was calculated by simply adding up the scores.

Respondents who shop exclusively in supermarkets, at conventional specialists or a combination of these two, rated sustainable alternatives less attractive than respondents who make use of alternative shops (exclusively or in combination with other shops) (Figure 4.11 and Table 4.5). Of respondents who shop exclusively in the supermarket on average 54% gave the rating attractive or very attractive to alternatives, and of those shopping at conventional specialists 58% gave the rating attractive/very attractive. In comparison, 72% of those shopping exclusively in alternative shops (and 65% of those shopping at alternative shops and the supermarket) gave attractive/very attractive ratings to alternatives. Respondents who shop at all three shops (supermarket, alternative shop and conventional specialist) are the most positive about sustainable food alternatives (76% of respondents).

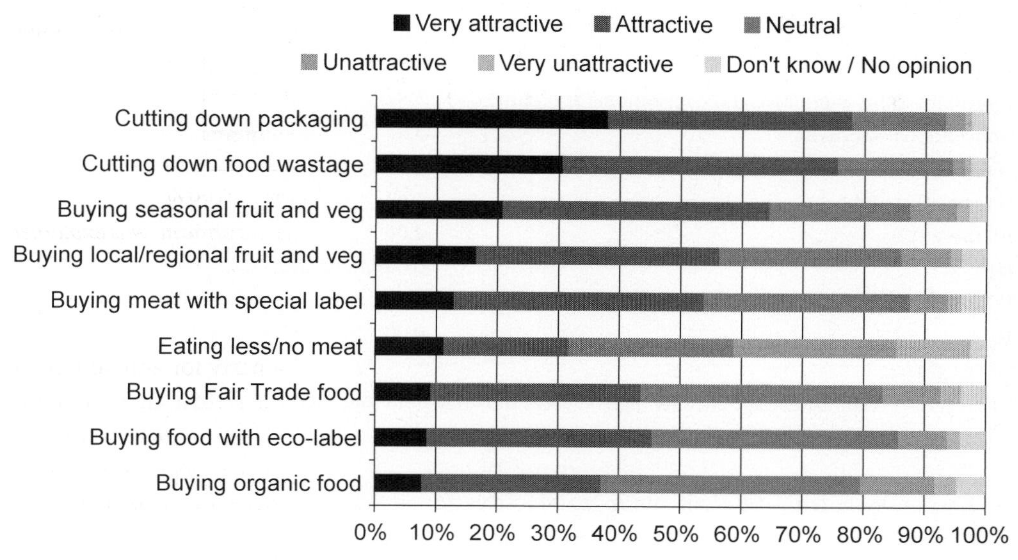


*Figure 4.10. Attractiveness of different food consumption alternatives.*

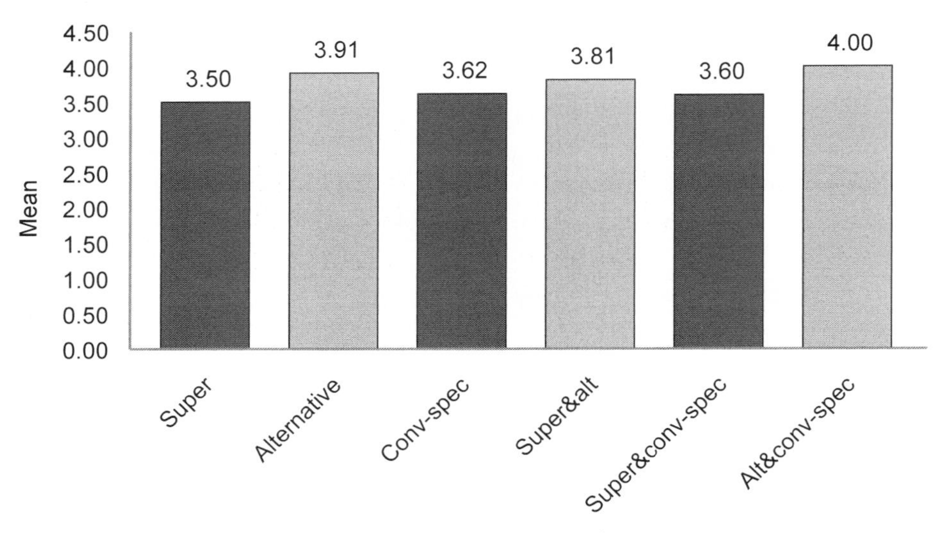


*Figure 4.11. Mean total scores of attractiveness ratings of different food consumption alternatives for respondents with different shopping practices (using 6 groups) (super = supermarket; conv-spec = conventional specialist) (scale: 1 = very unattractive, 2 = unattractive, 3 = neutral, 4 = attractive, 5 = very attractive).*

*Table 4.5. Mean scores of the total score on attractiveness of the different food alternatives.*

| | Mean[1] | Standard deviation |
|---|---|---|
| Supermarket | 3.50 | 0.59 |
| Alternative (plus alternative and conventional specialist) | **3.91** | 0.63 |
| Conventional specialist | 3.62 | 0.70 |
| Supermarket and alternative | **3.81** | 0.61 |
| Supermarket and conventional spec | 3.60 | 0.63 |
| All three | **4.00** | 0.58 |
| $Eta^2$ | 0.06 (moderate) | |
| F | 22.03 | |
| Sig. | 0.00 | |

[1] ANOVA, SNK post hoc: significantly higher scores shown in bold (Sig. 0.00).

Respondents with different shopping practices reveal similar patterns in which types of alternatives are found most and least attractive (Figure 4.12). For all shopping practices cutting down on food wastage, food packaging and eating seasonal produce are the top three most favourable alternatives (70-90% of respondents rated these alternatives as very attractive or attractive). Eating

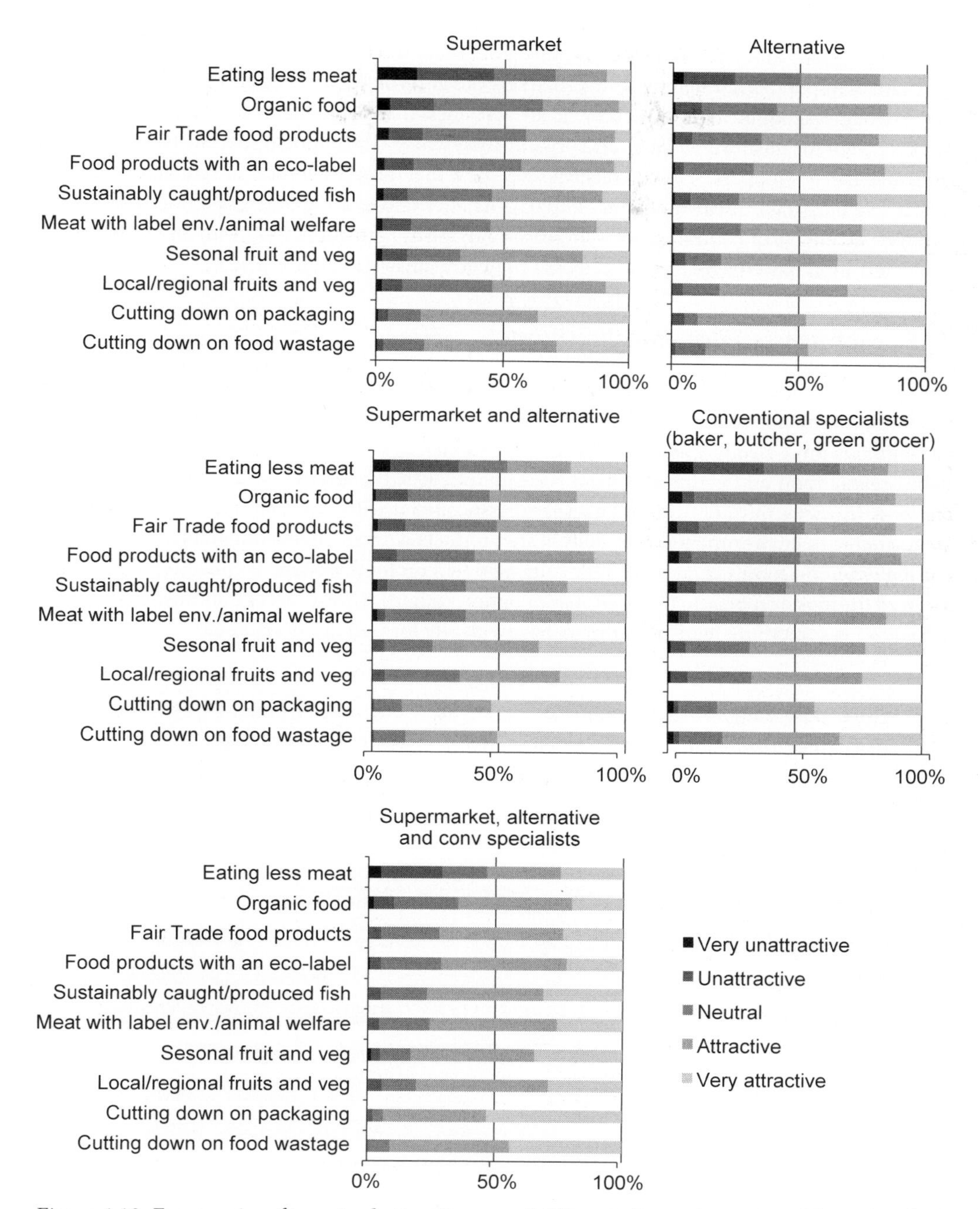


*Figure 4.12. Frequencies of perceived attractiveness of different alternatives as rated by respondents with different shopping practices.*

less meat, organic and Fair Trade food produces are the least attractive alternatives. For almost all alternatives respondents shopping at supermarkets and at conventional specialists (or a combination of the two) give a significantly lower score to attractiveness than respondents shopping at alternative shops (except for meat with a label, and organic food where conventional specialists do not differ from alternative shoppers). However, effect sizes show that these differences are small to moderate (effect size 0.02-0.05, small-moderate, $P<0.05$). As alternatives are found less attractive it seems that the percentage of neutral answers increases. Buying organic food, food with an eco-label and Fair Trade food have the highest percentages of neutral answers (Figure. 4.12).

### 4.7.3 Concerns associated with alternatives

The results on the concerns associated with each of the food alternatives gives some interesting insights into why consumers find these sustainable options appealing (Figure 4.13). Firstly, reducing packaging is most strongly associated with environmentally friendly behaviour (81%), more so than organic food and eating less meat for example. It is the alternative most associated with trustworthiness; a third of respondents (31%) associated it with this concern. Next in line to be associated with environmental friendliness are products with an environmental label (48%) and cutting down on food wastage (38.5%). Cutting down on food wastage is the alternative which received the most response to the open answer option ('other'). This seems to indicate that throwing away food is something many consumers have a clearly formed opinion about.

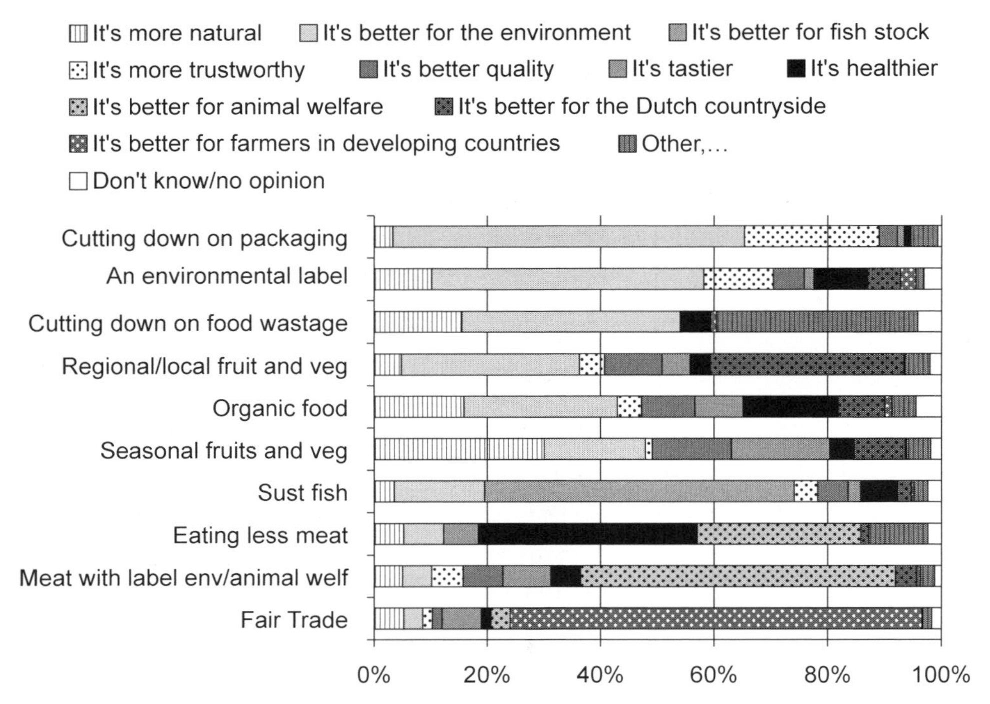


*Figure 4.13. Reasons respondents selected to explain why they find alternatives attractive.*

Secondly, for both meat with a special label and eating less meat about the same number of respondents make the association with environmental friendliness (5 and 7% respectively). Eating less or no meat is most associated with health benefits (39%) and animal friendliness (29%). Thus, whereas meat consumption is very much connected to (global) environmental impact, especially in relation to the greenhouse gas emissions and nutrient pollution, eating less meat or sustainably produced meat does not seem to be connected to a very strong environmental association. In fact, of all the alternatives eating less meat is the most associated with health (39%), whereas healthiness was not an often mentioned concern (on average 9% of respondents choose healthiness as a reason for finding an alternative attractive). As for animal friendliness, more respondents (55%) associate animal friendliness with special label meat (e.g. 'better life' labels, free range, organic, etc.) than with eating less meat (29%).

Thirdly, there are a number of surprising results on the associations with naturalness. Seasonal fruits and vegetables are associated the most with a more natural way of food consumption (30%) and less so with environmental friendliness (18%). In addition, almost just as many respondents associated cutting down on food wastage with naturalness as organic food (15-16%).

Fourthly, for many respondents better quality is not a very important association or reason to find an alternative attractive. In fact, better quality and taste are the least often chosen reasons for finding a food alternative attractive. Of respondents on average 7% consider better quality and 6.4% consider taste as a reason for finding a food alternative attractive.

## 4.8 Portfolio

### *4.8.1 Experience with sustainable alternatives*

#### *Frequency of use practical and product alternatives*

Making sure food is wasted as little as possible is something 83% of respondents do almost always or regularly, making it the most frequently engaged in alternative (Figure 4.14). Next in line are: buying seasonal fruits and vegetables (79% almost always or regularly), eating locally produced fruits and vegetables (61% almost always or regularly) and cutting down on packaging (fresh foods) (48% almost always or regularly). Consumers appear to have much less experience (on average 18%) with Fair Trade, animal/environmentally friendly meat, products with an environmental label and organic food products. Also, the percentage of respondents, who say they never make use of organic foods, meat with a sustainability label, sustainable fish, Fair Trade and compostable packaging is relatively high in comparison to the practical alternatives. Compostable packaging and sustainably farmed or caught fish are the least frequently used alternatives (69% of respondents never buy products packed in compostable packaging and 63% of respondents indicate never to buy sustainable fish).

Of respondents who indicated never to use a certain alternative (non-users) it was asked if they are familiar or unfamiliar with the alternative. Interestingly, respondents seem the least familiar with compostable packaging and sustainably farmed/caught fish (the least frequently used alternatives). More than half of respondents who never use compostable packaging and sustainable fish are also not familiar with these products (63.8% and 44% respectively) (Figure

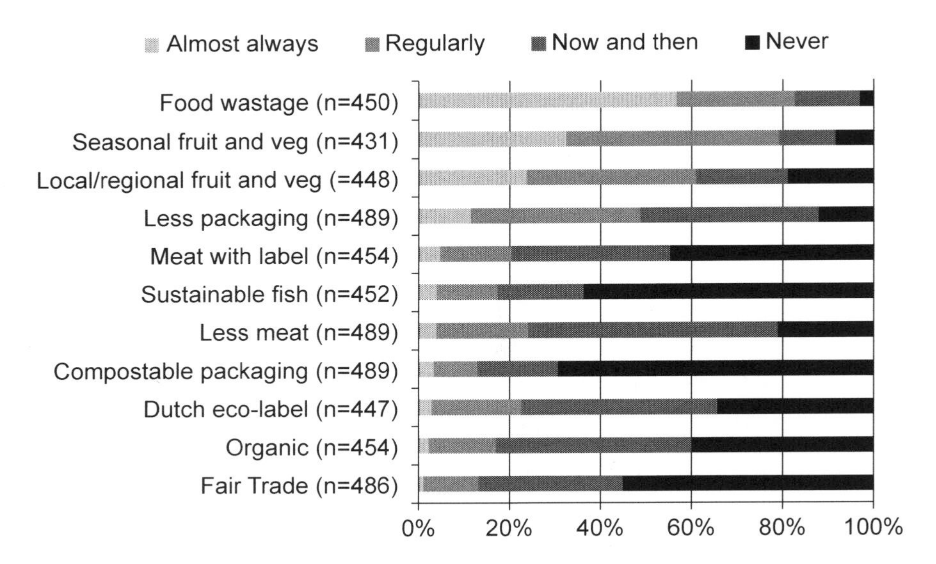


*Figure 4.14. Frequency of use of different sustainable food alternatives.*

4.15). Whereas many respondents indicate that they look to reduce the use of packaging on fresh produce, the use of compostable packaging (recognisable by a label on the packaging) is low. Most respondents who never use Fair Trade products, meat with an animal welfare/environmental care label, foods with an environmental label and organic are at least familiar with their existence.

As for the context in which respondents pay attention to sustainable alternatives, making a sustainable food choice is mainly an issue during grocery shopping (70-90% pay attention to sustainability during grocery shopping) (Figure 4.16). Respondents pay less attention to choosing sustainable food alternatives when eating out or eating lunch in work-place canteens (25%). In all three practices, the most attention is paid to food wastage, which is also the alternative most often engaged in as we saw previously with the experience results (Figure 4.14).

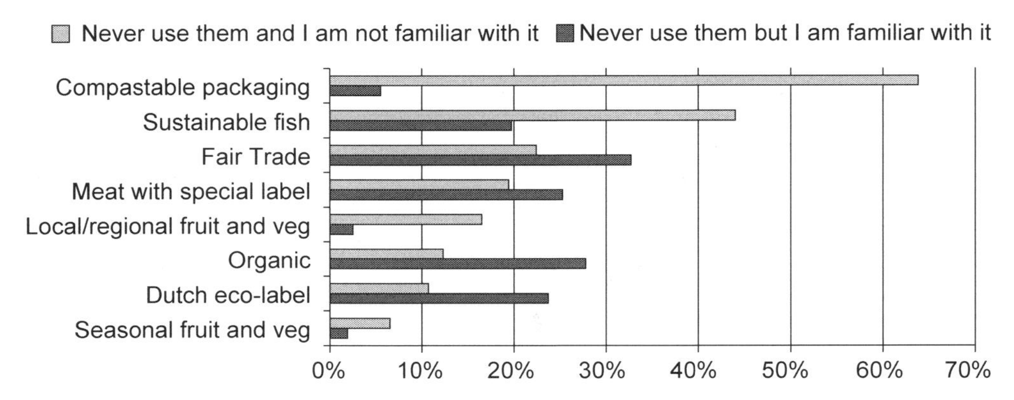


*Figure 4.15. Non-users and their familiarity with the alternative (n=431-489).*

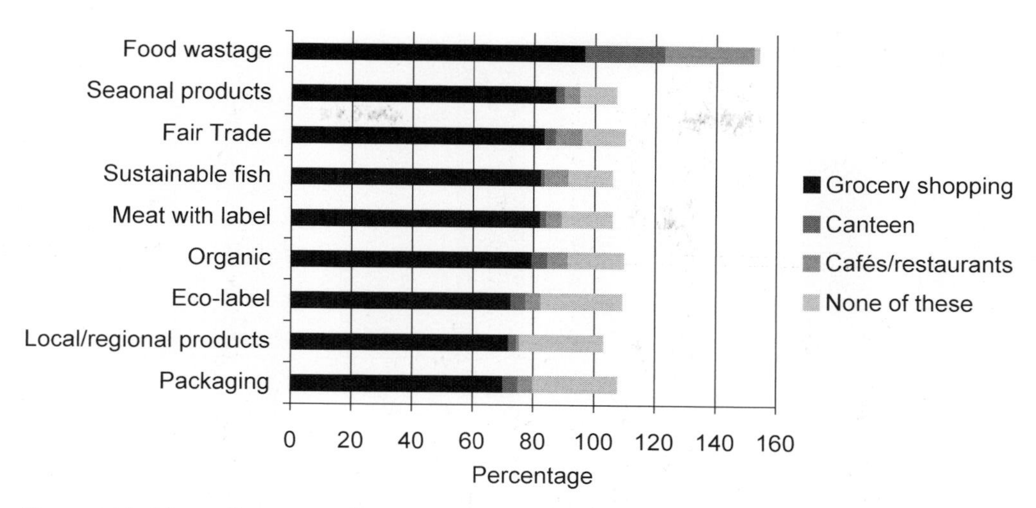


*Figure 4.16. Alternatives respondents pay attention to when grocery shopping, eating at work-place canteens and eating out in restaurants/café's.*

## *Shopping practices and the differences in use*

One-way ANOVA was carried out with a total score of frequency of use and the scores of frequency of use of each sustainable food alternatives separately. The first showed that respondents who shop exclusively in supermarkets make less frequent use of (are less experienced with) sustainable food alternatives than respondents who have other shopping practices (i.e. shop at alternatives hops, conventional specialists, or a combination of these shops) ($Eta^2$=0.06, moderate effect size, $P<0.05$) (Figure 4.17 and Table 4.6). There is no difference in total score between these other shopping practices, except for those shopping at all three shops, they have the highest experience score ($P<0.05$).

Analysis carried out on each alternative showed that respondents with different shopping practices differ the most markedly in their use of organic food, meat with a label for animal welfare and/or environmental care, Fair Trade products and food with a Dutch eco-label (large effect sizes, $P<0.05$, Table 4.7). Respondents who shop exclusively at supermarkets use these alternatives significantly less frequently than respondents who shop at alternative shops, be it exclusively or in combination with other types of shops. Except for seasonal fruits and vegetables, respondents who shop at alternative shops score lower than those who shop at the supermarket exclusively or in combination with conventional. For the practical alternatives using seasonal produce and eating less meat significant but small differences were found between shopping practices (small effect sizes, Table 4.7). Cutting down on food wastage is the only alternative which did not vary significantly between the different shopping practices ($P$=0.215).

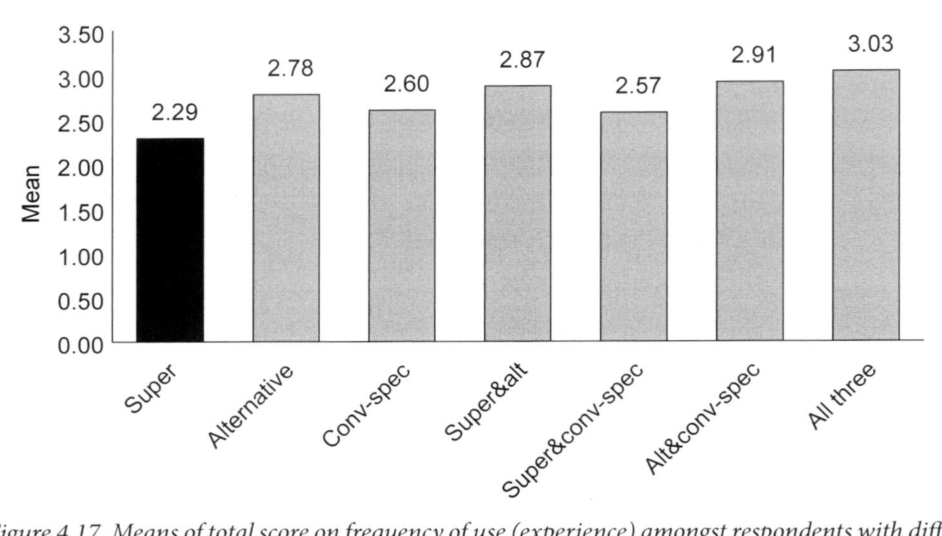


*Figure 4.17. Means of total score on frequency of use (experience) amongst respondents with different shopping practices (scale: 1 = never, 2 = now and again, 3 = regularly, 4 = almost always).*

*Table 4.6. Results from ANOVA and univariate analysis of variance on total frequency of use for respondents with different shopping.*

| | n | Mean[1] | Standard deviation |
|---|---|---|---|
| Supermarket | 825 | 2.28 | 0.93 |
| Alternative | 48 | 2.78 | 1.13 |
| Conventional specialist | 165 | 2.60 | 0.93 |
| Supermarket and alternative | 103 | 2.87 | 1.08 |
| Supermarket and conventional specialist | 426 | 2.57 | 0.96 |
| Alternative and conventional specialist | 69 | 2.91 | 0.97 |
| All three | 138 | 3.03 | 1.00 |
| $Eta^2$ | | 0.06 (moderate) | |
| F | | 19.67 | |
| Sig. | | 0.00 | |

[1] ANOVA, SNK post hoc.

*Table 4.7. Overview of results from ANOVA and analysis of variance (univariate) on the differences between shopping practices in terms of use of different sustainable food alternatives.*[1,2]

| **Product alternatives** | **Organic (n=454)** | **Meat with label (n=454)** | **Sustainable fish (n=452)** | **Fair Trade (n=482)** | **Compostable packaging (n=498)** | **Environmental/ Eco-label (n=447)** |
|---|---|---|---|---|---|---|
| ANOVA Post hoc SNK (summary) | alt > super; alt-conv-spec > all others | alt > super | none | alt > super | none | Alt > super; all three > others |
| F | 10.19 | 10.39 | 9.10 | 14.76 | 4.23 | 10.46 |
| Eta$^2$ | 0.15 (large) | 0.15 (large) | 0.09 (moderate-large) | 0.18 (large) | 0.06 (moderate) | 0.15 (large) |
| Sig. | **0.000** | **0.000** | **0.000** | **0.000** | **0.000** | **0.000** |

| **Practical alternatives** | **Seasonal (n=431)** | **Cutting down on packaging (n=430)** | **Eating less meat (n=489)** | **Regional/local (n=448)** | **Cutting down on food wastage (n=450)** |
|---|---|---|---|---|---|
| ANOVA Post hoc SNK (summary) | super-alt, alt-conv-spec, all three > alt, super, super-conv-spec | SNK: no differences between groups | alt > super | alt, all three > super | none |
| F | 2.81 | 4.66 | 4.85 | 7.34 | 1.63 |
| Eta$^2$ | 0.05 (small-moderate) | 0.06 (moderate) | 0.03 (small-moderate) | 0.11 (moderate-large) | - |
| Sig. | **0.016** | **0.000** | **0.000** | **0.00** | 0.139 |

[1] Alt = alternative shop; conv-spec = conventional specialist shop; super = supermarket; alt-conv-spec = alternative and conventional specialist shops; super-alt = supermarket and alternative shop; super-conv-spec = supermarket and conventional specialist shop.

[2] Significantly higher results are shown in bold.

## *4.8.2 Food knowledge*

More respondents seem knowledgeable about the major agro-food issues of our time than on how to make their own food habits more sustainable. On average, 51% of respondents said they knew 'something' or 'a lot' about issues relating to unsustainable fishing, environmental problems caused by the agro-food industry and problems relating to trade issues. In comparison, only 20-30% of all respondents agreed they knew 'something' or 'a lot' about which food products are sustainable and which are not, and how to make their meals more sustainable (Figure 4.18).

Overall, differences in the total food knowledge score (reliable Cronbach's alpha, Table 4.8) between respondents with different shopping practice were highly significant and moderate to large (effect size = 0.07, $P<0.05$, Table 4.8). Respondents who shop at alternatives shops, be it exclusively or in combination with other shops, show they know more about food sustainability issues than respondents who shop at the supermarket (Figure 4.19).

### *4.8.3 Use of information*

Of all the different sources of information on food sustainability respondents' own judgement, knowledge and experience scores as the most frequent (Figure 4.20). This result is supported by other studies which show that consumer choices are made on the basis of a pool of personal know-how and experience (Aubrun *et al.*, 2005; Bunte *et al.*, 2008; INSnet, 2007; Wertheim-Heck, 2005). The INSnet Monitor 2007 for instance found that for the daily purchase of sustainable alternatives (examples given were toilet paper, bulbs or batteries) 57% of respondents said they base their choice of products on their own judgment and 42% on past experience. Similarly, Bunte *et al.* (2008) found that knowledge plays an important role in people's food choice, but that this knowledge is based more on opinion than factual information.

On-site information (shop-floor information and information from shop employees), as well as information from consumer organisations are the least consulted (11% and 6%, and 10% respectively, consult these regularly or often). Again, the results from the INSnet Monitor 2007

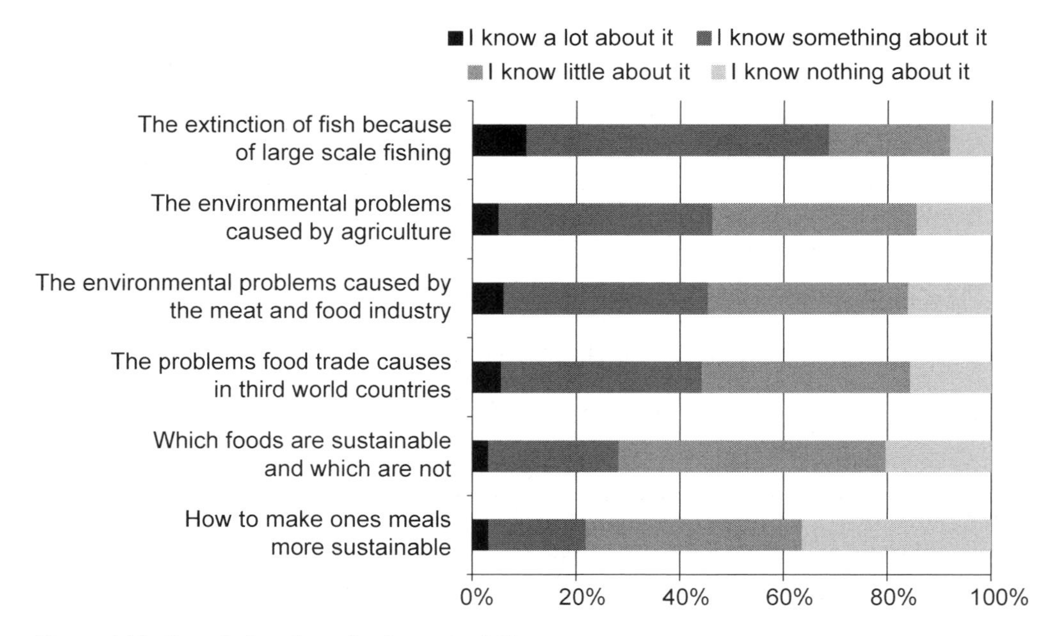


*Figure 4.18. Knowledge about food sustainability.*

*Table 4.8. Overview of results from ANOVA and analysis of variance (univariate) on the differences between shopping practices in terms of food knowledge.*

| | n | Mean[1] | Standard deviation |
|---|---|---|---|
| Supermarket | 846 | 2.19 | 0.59 |
| Alternative | 48 | **2.75** | 0.62 |
| Conventional specialist | 172 | 2.43 | 0.59 |
| Supermarket and alternative | 106 | **2.57** | 0.66 |
| Supermarket and conventional specialist | 434 | **2.34** | 0.59 |
| Alternative and conventional specialist | 71 | **2.60** | 0.53 |
| All three | 143 | **2.57** | 0.61 |
| Cronbach's alpha | | 0.90 | |
| Univariate AoV $Eta^2$ | | 0.07 (moderate-large) | |
| F | | 21.65 | |
| Sig. | | 0.000 | |

[1] SNK post hoc: significantly higher scores are shown in bold (Sig.=0.00).

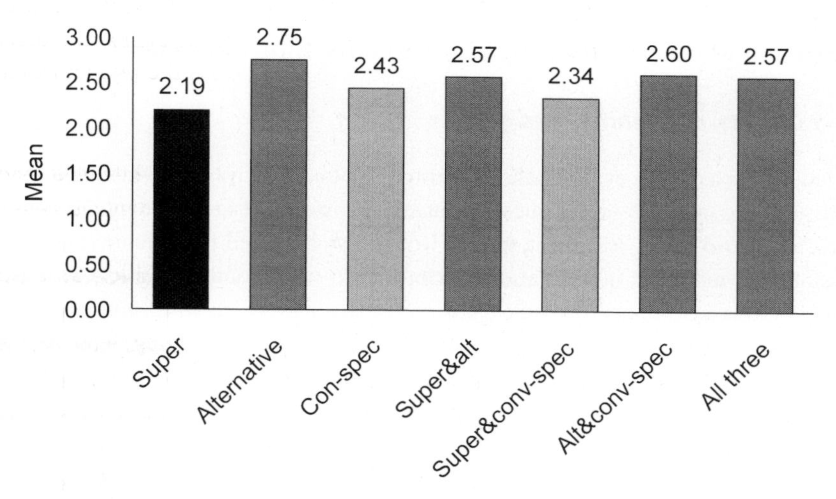


*Figure 4.19. Food knowledge of respondents who shop at different shopping channels. (scale: 1 = don't know anything about it, 2 = know little about it, 3 = know something about it, 4 = know a lot about it).*

support this. It found that respondents prefer extra information if it comes from environmental organisations (NGO's) and not from shops themselves or shop-keepers. Information from providers (retailers) and shop staff is not something which consumers consider very trustworthy.

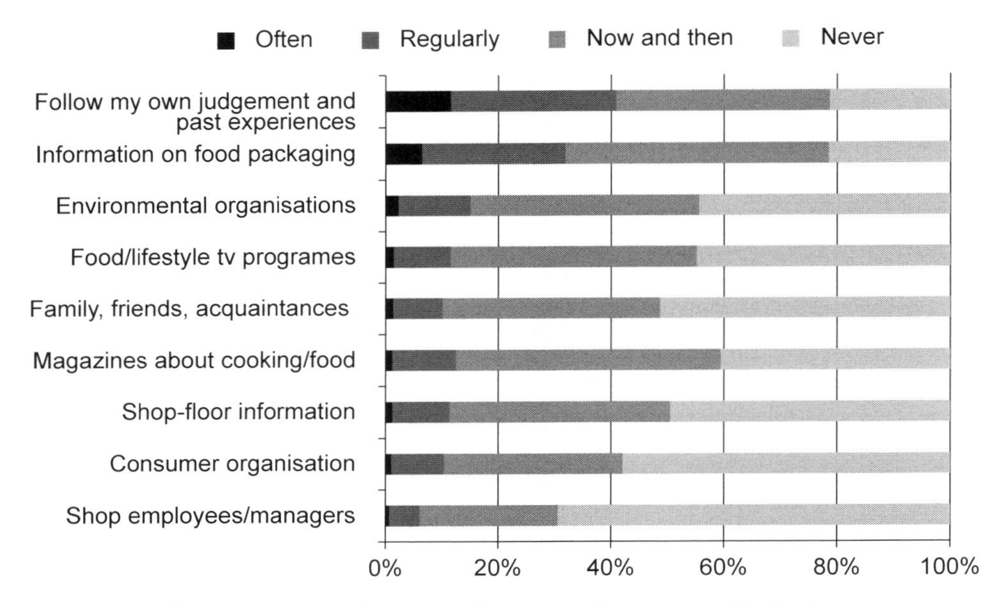


*Figure 4.20. Different ways respondents get information about sustainable food.*

## 4.9 Evaluation of the provisioning of sustainable food alternatives

### *4.9.1 Accessibility, availability and quality*

Respondents were asked to evaluate the provisioning of alternatives for which they had already answered questions on portfolio (frequency of use). For instance, if a respondent received questions on organic food and eating less meat, he/she would also be asked to evaluate these alternatives on accessibility, availability, quality, and contribution to sustainability (Table 4.5). Respondents receiving questions on cutting down on packaging and use of products with compostable packaging were given the same evaluation questions (overall evaluation on packaging).

For each alternative, questions were carefully selected to come to an evaluation of the various aspects (Table 4.9). This means that not exactly the same questions were used for every alternative. For example for 'eating less meat' respondents were asked if they liked eating meat-replacements and vegetarian food in order to judge the accessibility of this alternative (Appendix 1).

Figure 4.21 shows the percentage of respondents that agree or strongly agree with the statements on the degree of accessibility, availability, quality and contribution to sustainability of each of the sustainable food alternatives. Product alternatives received a positive evaluation by 30% to 40% of respondents, whereas for all practical alternatives more than 40% of respondents gave a positive evaluation.

Fair Trade foods appear to be the most available foods (33%) (Figure 4.22). In comparison around 25% of respondents agree or strongly agree that the other product alternatives are readily available. Of the product alternatives organic foods are considered the most accessible (43%), and fish with a sustainability label the least accessible of the product alternatives (25%). In comparison

*Table 4.9. Types of questions use in the evaluation of provisioning of different sustainable food alternatives.*

| | **Type of question** |
|---|---|
| Accessibility | Is the alternative easily accessible: is it easy to get it/find it in shops; is it easy to use/do; is it tasty |
| Availability | Is there enough choice (range of products available) |
| Quality | Is the quality good and/or is it user friendly and/or is it trustworthy |
| Sustainability effect | Is the alternative a good way to contribute to sustainable food consumption |

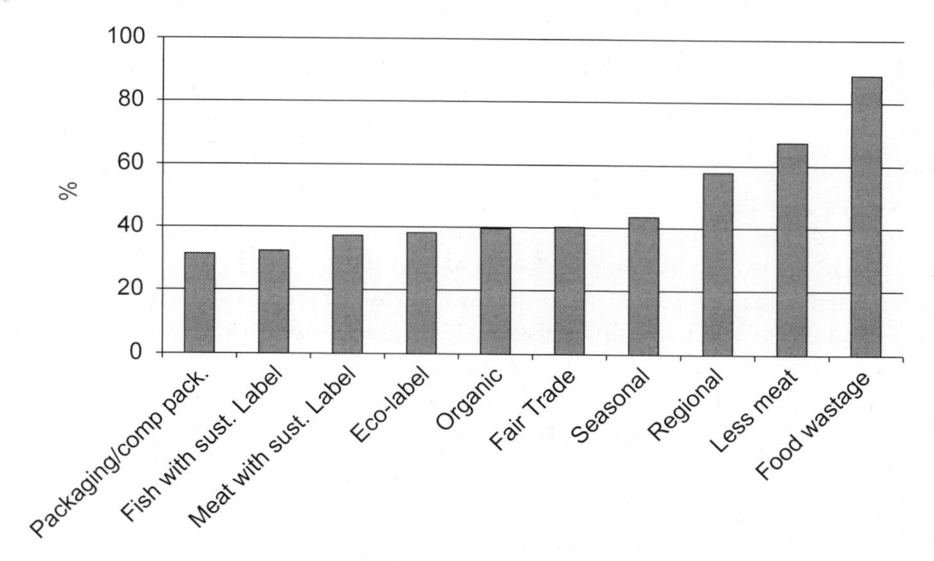


*Figure 4.21. Percentage of respondents giving a positive evaluation to various sustainable alternatives (percentage that agree or strongly agree with statements on the degree of accessibility, availability, quality and contribution to sustainability of sustainable alternatives).*

cutting down food wastage, seasonal foods and regional/local produce were considered available and accessible by near to 60% of respondents (Figure 4.23).

### *4.9.2 Sustainability effect*

Comparing the practical alternatives with each other, and with the product alternatives, is less straightforward because of the different kinds of questions posed for their evaluation. However, we can do some comparison on the basis of the sustainability effect and accessibility (easy to get hold of/easy to use/easy to do) of these alternatives. 'Cutting down on food wastage' appears to

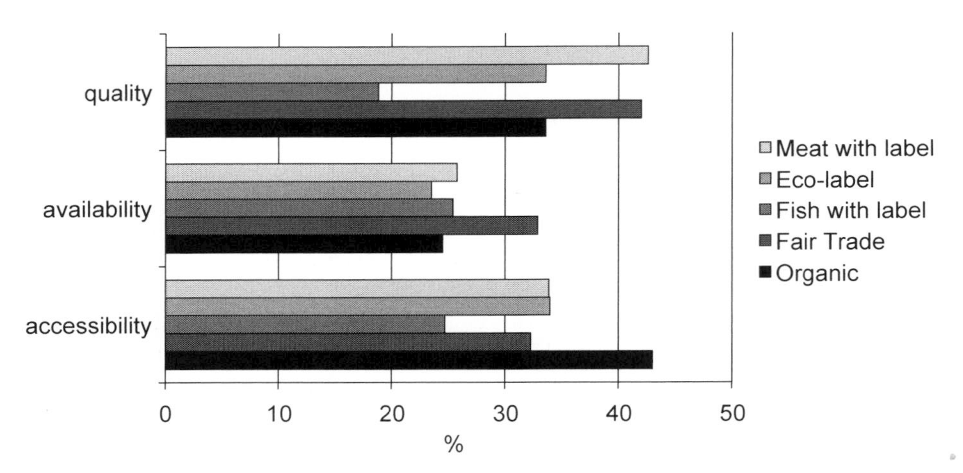


*Figure 4.22. Evaluation of the provisioning of product alternatives (percentage of respondents answering agree and strongly agree).*

be most often associated with a good way to contribute to sustainability and seems to be the most accessible: 66% of respondents agree/totally agree that this is a good way of doing something for the environment, and 72% agree/strongly agree that it is 'it's easy to do'. Also, 66% of respondents agreed/strongly agreed that 'I'm used to doing it' (Figure 4.23).

Paying attention to packaging[32] (60%) and buying seasonal produce (58%) are next in line in terms of sustainability effect. In terms of accessibility, seasonal produce (61%) scores higher than paying attention to packaging (average of 38%; covers cutting down on packaging and buying products with compostable packaging). For packaging 60% of respondents indicate they can easily get hold of fresh fruits and vegetables without packaging but only 16% indicate they can easily get hold of products with compostable packaging around them.

Eating less meat is the least associated with a viable sustainability effect (36%). Although around half of respondents indicate they have access to alternatives to meat (55%), only around 24% say they like meat replacements, which seriously demeans this alternative's accessibility. Even fewer respondents indicate that vegetarian style cooking suits them (15%) or that they like vegetarian food (27%).

Overall, both product and practical alternatives received a similar evaluation for sustainability effect. That is, for both product and practical alternatives a similar number of respondents evaluated sustainability effect positively (around or near to 60% of respondents for each alternative), with the exception of Fair Trade food products (51%), eating less meat (36%) and cutting down on food wastage (66%) (Figure 4.24).

[32] Respondents who had previously answered questions on 'cutting down on packaging' or on 'compostable packaging' received the same evaluation questions, referred to as 'paying attention to packaging', see Appendix 1.

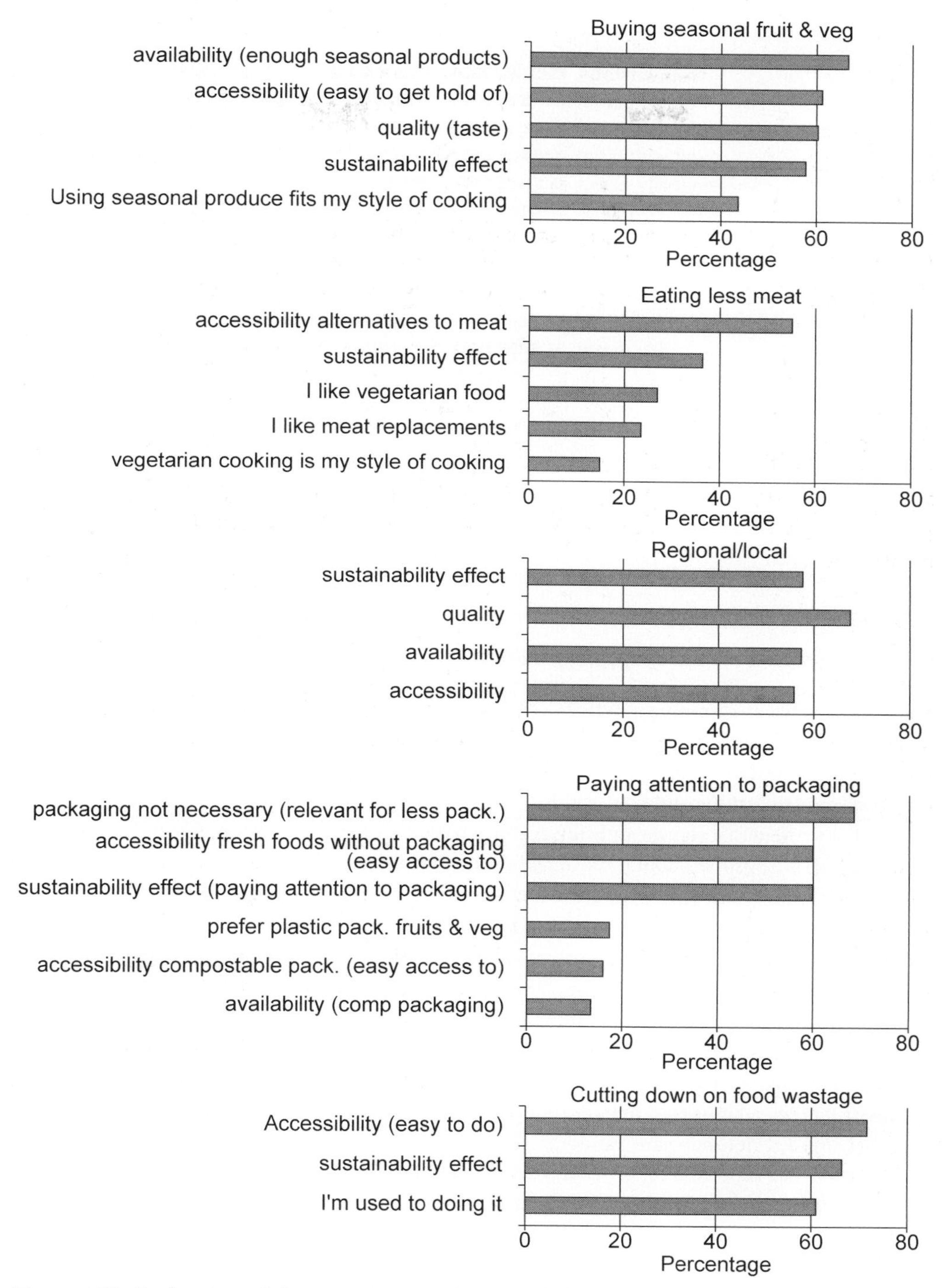


*Figure 4.23. Evaluation of the provisioning of practical alternatives (percentage of respondents answering agree and strongly agree).*

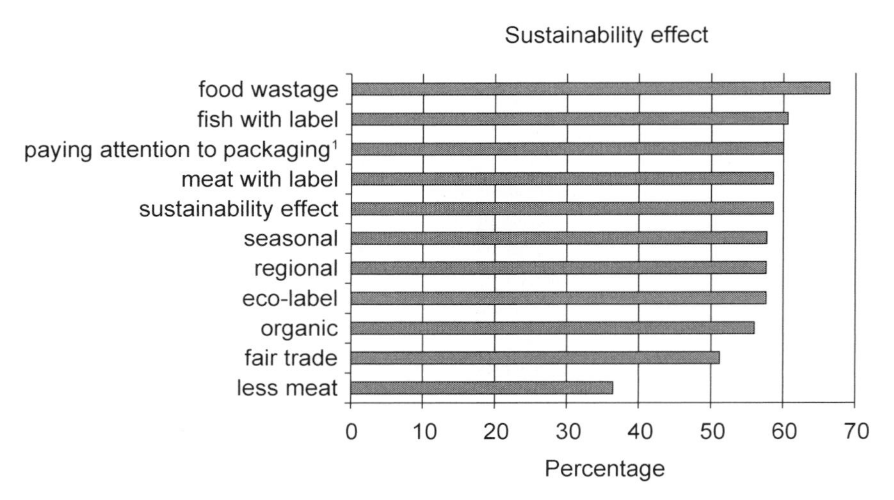


*Figure 4.24. Sustainability effect of the various food alternatives showing percentages of respondents answering agree or strongly agree.*
*[1] Paying attention to packaging included questions on both cutting down on packaging and compostable packaging.*

### *4.9.3 Differences between shopping practices*

Differences between shopping practices were measured in two ways, by looking at the total ratings of availability, accessibility, quality and sustainability effect[33] and by considering the evaluations given to each alternative. Respondents were given the same alternatives to evaluate as those which they answered questions on for (portfolio) frequency of use.

Analyses reveal that significantly fewer supermarket visitors (those shopping exclusively at the supermarket) are positive about the availability and the accessibility of sustainable food alternatives compared with respondents with any other shopping practices, though these differences were small-moderate (Table 4.10). Similarly, significantly fewer supermarket shoppers are positive about the quality of sustainable food alternatives compared to alternative shoppers and those shopping exclusively at conventional specialists. The opinion on the sustainability effect of alternatives hardly differed between the various shopping practices.

Scores on availability, accessibility, quality and sustainability effect were totalled per alternative, and this score was then used in comparing groups (6 groups). Post hoc tests show that significant, effective differences between groups are only found for seasonal, organic and eco-labelled products, and that these differences are moderate-large ($P<0.05$, Table 4.11). Respondents who shop exclusively in the supermarket give a less positive evaluation of the provisioning of seasonal food than respondents who also shop in alternative shops and the supermarket, or those who shop in all three shops (supermarket, alterative shops and conventional specialists) (moderate-large

[33] Scores on each of these across alternatives were totalled (i.e. the total score of accessibility for all alternatives, total score of availability of all alternatives, etc.).

*Table 4.10. Overview results from analysis of variance (univariate) for the evaluation of provisioning: differences between consumers with different shopping practices (6 groups).*[1]

| Evaluation provisioning | Availability | Accessibility | Quality | Sustainability effect |
|---|---|---|---|---|
| ANOVA Post hoc SNK (summary) | super < other shopping practices | super < other shopping practices | super < those who shop at alt and conv-spec shops | none |
| Effect size Eta$^2$ | 0.019 (small-moderate) | 0.032 (small-moderate) | 0.019 (small-moderate) | 0.013 (small) |
| F | 7.120 | 12.067 | 6.987 | 4.923 |
| Sig. | 0.00 | 0.00 | 0.00 | 0.00 |

[1] Alt = alternative shop; conv-spec = conventional specialist shop; super = supermarket.

*Table 4.11 Overview results from ANOVA and analysis of variance (univariate) for the evaluation of the product alternatives (6 groups).*[1,2]

| Product alternatives | Organic (n=454) | Regional/local (n=448) | Meat with label (n=454) | Sustainable fish (n=452) | Fair Trade (n=482) | Eco-label (n=447) |
|---|---|---|---|---|---|---|
| ANOVA Post hoc SNK (summary) | all three shops > other shopping practices | none | none | none | none | all three shops > other shopping practices |
| F | 4.378 | 2.034 | 2.033 | 1.151 | 2.483 | 4.092 |
| Eta$^2$ | 0.07 (moderate-large) | 0.03 (small-moderate) | 0.04 (small-moderate) | 0.03 (small-moderate) | 0.04 (small-moderate) | 0.07 (moderate-large) |
| Sig. | **0.001** | 0.074 | 0.074 | 0.335 | **0.032** | **0.001** |

| Practical alternatives | Seasonal (n=431) | Packaging[3] (n=430) | Eating less meat (n=489) | Cutting down on food wastage (n=450) |
|---|---|---|---|---|
| ANOVA Post hoc SNK (summary) | Super < super-alt, all three shops | none | none | none |
| F | 4.893 | 0.607 | 2.823 | 1.842 |
| Eta$^2$ | **0.08** (moderate-large) | 0.01 (small) | 0.05 (small-moderate) | 0.03 (small-moderate) |
| Sig. | 0.000 | 0.626 | **0.017** | 0.104 |

[1] Super = supermarket; super-alt = supermarket and alternative shop.

[2] Bold values are significant.

[3] Cutting down on packaging and use of products with compostable packaging is taken together here as the question on packaging covers both alternatives).

difference). Respondents who shop in all three shops are most positive about the provisioning of organic and eco-labelled food (moderate-large) compared to the other shopping practices.

## 4.10 Discussion

### *4.10.1 How do Dutch consumers view their role in sustainable development of food consumption and to what extent do they engage in a range of different sustainable food alternatives?*

*Consumer co-responsibility*

Although around half of Dutch consumers understand the necessity for more sustainable food consumption patterns and recognise consumers' responsibility to play a part in this, they seem less sure/opinionated about the active role consumers should and/or can take. Also, they appear less positive about their own active contribution. The importance of sustainable food consumption seems well established when it comes to perception on the general discourse on sustainable development and food, but less well established when it comes sustainability in the context of daily consumption practice(s). There are large percentages of neutral answers on people's own active role. This may indicate the sporadic, unroutinised nature of sustainable consumption practice under Dutch consumers. Alternatively, it might indicate a certain degree of ambivalence when taken together with the finding that Dutch consumers seem unsure as to whether conventional food products are already sustainable enough (high percentage of 'I don't know's').[34] The findings on co-responsibility reveal the different facets of co-responsibility and how the issue of sustainable food consumption might be experienced by consumers within the context of everyday practice. It also suggests the ambiguous nature of the term 'consumer responsibility', which can be viewed in different ways; consumers may find that sustainable consumption in general is an important issue and thus support it, but also that they themselves cannot judge which foods are sustainable and which are not. Or they may simply not be prepared or able to actively seek out more sustainable foods. Others may indeed be prepared and able to make more sustainable consumer choices, but may do so for other reasons than sustainability. The question is then, which consumer is to be classed as 'responsible'? The consumer who 'acts responsible' but does not 'feel responsible', or vice versa?

*Sustainable food alternatives: their attractiveness, use and evaluation in everyday food practice*

Overall, practical alternatives (with the exception of eating less meat) are not only considered to be relatively more attractive than product alternatives (like organic food), they are also engaged in more often. It seems that the most used alternatives in sustainable daily food practice are cutting

[34] It should be noted that the statement 'Conventional products are sustainable enough' on which respondents were asked to comment was maybe too vague which may have left some respondents confused as to what to answer.

down on food wastage (83%) and buying seasonal fruits and vegetables (79%). The least often used alternatives are organic foods, Fair Trade foods, and food product with the Dutch eco-label (between 15-20%). Overall, the alternatives which are the least frequently used are also found to be the least available and the least accessible. Overall, practical alternatives received higher ratings on accessibility (>55%) than most of the product alternatives (organic, Fair Trade, meat with label, fish with label, eco-label) (<40%). Regional/locally produced fruits and vegetables are considered to be the most available and accessible, whereas organic food, sustainable fish and food products with an eco-label are considered the least available, and Fair Trade foods and sustainable fish are the least accessible. The sustainability effect of all alternatives are rated similarly (with the exception of eating less meat).

Not only do more respondents find product alternatives like Fair Trade and organic food products relatively less attractive to engage in, many consumers seem undecided on these products; the percentage of 'neutral' answers more or less increases with decreasing popularity (whereas the percentage unattractive/very unattractive remains relatively low). Although fair trade is a top consumer concern, Fair Trade products are rated attractive/very attractive by substantially less respondents (45%) than for instance cutting down on packaging (80% of respondents).

The findings suggest that a number of practical sustainable alternatives (cutting down on food wastage, eating seasonal and locally/regionally produced foods) are well-established alternatives within Dutch food consumption practices whereas the use of products like organic food, Fair Trade labelled food products and eating less meat are not. Similar patterns are found for the attractiveness of alternatives and their accessibility and availability. In terms of everyday food practice, this suggests that widely used alternatives are those which do not require substantial changes in cooking, eating and food shopping practices in terms of switching to other products, or to other shops, more/temporary investment in terms of cooking skills, time and money, etc. For instance, eating less meat is engaged in the least and is considered the least attractive. This alternative can require fairly drastic changes in daily shopping and cooking practices and altering deeply rooted traditions and the food culture surrounding meat. Although one might think that activities like cutting down on packaging, wasting as little food as possible and buying local/seasonal foods could be perceived by consumers as more tangible and effective ways of 'engaging in sustainable consumption', the findings do not support this. The perceived sustainability effect of all the food alternatives are fairly similar, also when we look at respondents with different shopping practices. This suggests that other issues and considerations are playing a role here.

The findings on peoples' associations with different alternatives gave some surprising insights. Reducing packaging was most associated with environmentally friendliness. In comparison, eating less meat for instance was not often associated with environmental benefit. Taste and quality were two of the least often chosen associations linked to sustainable food alternatives. Since these aspects are often mentioned as important factors in food choice one would have expected them to be chosen more frequently. However, when it comes to sustainable food alternatives other associations such as environmental friendliness and naturalness appear to be more dominant. This might point to the fact that many Dutch consumers have an understanding of these alternatives in terms of their 'abstract' sustainability qualities, but less so in terms of their 'use-qualities' (their taste and quality; nutrition wise, culinary wise) which are based on actual experience with these products (cooking and tasting them).

*Knowledge and use of information*

The findings show that more respondents are knowledgeable about a number of important agro-food sustainability issues ('factual knowledge') than about which foods are sustainable and how to make their meals more sustainable. This suggests that the general awareness and understanding of sustainability issues related to food is generally greater than the understanding of how to realise sustainability within everyday food practice. Similarly, Tanner & Kast (2003) found that action-related knowledge is a determining factor of the purchase of sustainable foods. The findings suggests that consumers appear to make use of personal rules-of-thumb based on own established routine, frames of reference and experience. This appears everything but a fact-based and structured way of decision making. The findings show that the use of shop-floor information (which includes information from shop employees and information present within the shop) is not a substantial part of consumer's information sources on sustainable food products. However, consumers might not be consciously aware of how 'on-site' information, shop and shelf design influences their food practices. Such factors have been shown to play a role in food choice (see for instance the experiment by Dagevos *et al.*, 2005).[35]

### 4.10.2 How do consumers with different shopping practices differ in their engagement in sustainable food consumption?

One might expect respondents who shop in alternative shops to score much higher on co-responsibility (total score), but although significant, the difference in score between those who shop at alternative shops and those who shop at supermarkets is marginal. Furthermore, respondents with different shopping practices do not score differently on their opinions of the link between the environment and food consumption. However, the differences between respondents with different shopping practices becomes more marked when the subject is applying sustainability in practice; those who shop exclusively in supermarkets are less inclined to make sure the food they buy is sustainable compared with respondents with other kinds of shopping practices (statement: 'I make sure the food I buy is sustainable'). Thus, supermarket shoppers appear to be not less responsible overall, but are less likely to come home with a sustainable alternative.

When we look at the result of consumer portfolio we see a similar picture. It seems that overall respondents who shop exclusively at supermarkets have less knowledge and experience with sustainable food consumption compared to respondents who (also) make use of alternative shops (organic/natural food shops/farmers shops/markets). The most pronounced differences between shopping practices of the entire survey are found in respondents' use of certain product alternatives: people who shop exclusively in the supermarket make less frequent use of regional/local, sustainably produced fish, organic, eco-labelled and Fair Trade food, and meat with a label on animal/environmental care. Except for buying seasonal fruits and vegetables; this is practiced equally by those visiting supermarkets and alternatives shops. All respondents appear to engage similarly often in cutting down on food wastage and packaging. Thus, a general conclusion might

[35] And certainly shop-floor design and marketing are already used, only maybe not to encourage sustainable choices specifically.

be; consumers who shop at supermarkets have a less well developed portfolio for sustainable food consumption in terms of sustainability knowledge and the use of product alternatives.

Respondents who shop exclusively in supermarkets, at conventional specialists or a combination of these two, rated sustainable alternatives less attractive than respondents who make use of alternative shops (exclusively or in combination with other shops) (except for meat with a label, and organic food where conventional specialists do not differ from alternative shoppers). However, effect sizes show that these differences are small to moderate. Respondents with different shopping practices do reveal similar patterns in which types of alternatives are found most and least attractive; cutting down on food wastage, food packaging and eating seasonal produce are the top three most attractive. Eating less meat, organic and Fair Trade food produces are the least attractive alternatives. As alternatives are found less attractive it seems that the percentage of neutral answers increases (organic food, food with an eco-label and Fair Trade food have the highest percentages of neutral answers). This suggests the undecided attitude many respondents have in relation to these products.

Generally, supermarket shoppers give the most negative ratings on the accessibility, availability and quality of sustainable food alternatives, although these differences are not very substantial. When we consider the evaluation of each alternative separately, those shopping at alternative shops do not give a significantly higher positive rating to alternatives. Only organic foods, eco-labelled foods and seasonal produce stands out in revealing substantially significant differences and here it is those respondents shopping at all three shops who are the only group to give more positive evaluations.

Overall, the findings illustrate that consumers with different shopping practices have different portfolios, that is, different knowledge about food sustainability and/or experience(s) with sustainable food alternatives. The differences are found especially in terms of use of certain product alternatives. Portfolios are much less different in terms of the use of practical alternatives like cutting down on food wastage and packaging and the degree to which alternatives are positively evaluated.

### *4.10.3 What can we learn from the use of the SPA framework for a quantitative survey as carried out in this study?*

Using the SPA framework has given insight into how Dutch consumers engage in everyday sustainable food consumption and shows how consumers with different shopping practices differ from each other in this engagement, in terms of their knowledge about and experience with sustainable food alternatives. However, the survey reveals the need to further investigate food consumption in relation to practices of food consumption and their contexts. It seems that sustainable alternatives are substantially more often considered within the context of grocery shopping compared with canteen lunch practices or eating out (Section 4.8.1). This begs the question whether the results presented in this chapter would have been similar had the entire survey been set in the context of 'eating outside the home' or some other social practice? Understanding sustainable food choice and perceptions can really only become more representative when they are placed in the context of where and when. In this respect the survey could have done more justice to the SPA framework, more firmly anchoring it in context-specificity.

For further research, it would be interesting to investigate how different social practices represent different forms of contact and interaction with different 'consumption infrastructures' and/or how this determines and shapes food concerns, portfolios, etc. For instance, those making use of supermarkets will encounter a different sustainable provisioning infrastructure compared to those who shop at alternative shops. Do these also represent different sets of/frames of references, associations, expectations and standards (in relation to sustainability, healthy eating, cooking, food quality taste and more)? Besides understanding what is happening in the household sphere, it would also be interesting to understand what is happening in out-of-home food consumption and the importance that 'food services consumption' has on shaping daily eating habits. An interesting, largely unresearched area of study concerns cooking and food preparation. Questions posed might be: What is the state of cooking and food preparation skills in the Netherlands? Do people have the skills to prepare a 'sustainable' and/or healthy meal on a daily basis? What is the role of food providers in this? How do consumers with different food shopping practices differ in their cooking and food preparation practices and skills?

# Chapter 5. Sustainable catering: realising sustainable canteen provisioning

## 5.1 Introduction

Various studies have indicated the importance of the catering sector and food procurement for sustainable development of food consumption (Mikkelsen *et al.*, 2005; Mikkelsen *et al.*, 2002; Morgan, 2008). Targeting the food procurement of the 'public plate' (Morgan, 2008) as well as private businesses is seen as a way to boost the sustainable food economy (DEFRA, 2006; EC, 2004b; Eurocities, 2005) and improve consumers' daily access to healthy and sustainable produced foods. However, despite its potential this remains largely unexplored terrain. For instance, Dutch NGO's have not targeted the catering sector in the same way as they challenge ('name and shame') supermarkets.[36] This despite the fact that caterers are powerful players in the food services industry, in charge of public and private food services with an enormous reach. They influence the daily eating habits of people of all ages and all walks of life; school children, students, patients, prison inmates, travellers and employees.

The relevance of the catering sector should also be seen in light of the growing need within the society for (better) food services (Roos *et al.*, 2004) and (un)healthy eating patterns. Many consumers welcome convenient ways to meet their food requirements within their busy lifestyles; see for instance the growth convenient shopping solutions (e.g. on-line food shops), ready-meals and 'food to go' products (Gfk, 2008; Jabs *et al.*, 2007). Unhealthy eating habits and obesity is a growing societal problem, especially worrying are the figures on the unhealthy eating and drinking habits of youngsters and children (1 in 7 youths aged 2-24 is overweight; CBS, 2007) (Jeugdraad, 2008; Tacken *et al.*, 2010). Unhealthy eating habits amongst adults and youngsters has been connected to the intake of high-energy foods, unhealthy fats and the lack of fruit and vegetables. The latter has been identified as one of the most important health issues (Kreijl, 2004; Westert *et al.*, 2008). Especially individuals with a low socio-economic status have a low intake of fresh fruits and vegetables (Jansen *et al.*, 2002; Kreijl, 2004). Stimulating the consumption of fruit and vegetables amongst certain groups would decrease risk of heart disease, cancer, diabetes and obesity. Various studies from other European countries support this.[37] In such a context, catering services could play a huge role in improving diets, especially those of young people. [38]

[36] This on issues such as pesticide residues and the meat industry. For example, the Dutch version of Friends of the Earth (Milieudefensie) has the following campaigns: the 'Know what you're eating' campaign ('Weet wat je eet').

[37] For example, Bowman (2006) and Erkkilä (1999) show that there is a link between low socio-economic status and dietary linked coronary heart disease. The 2007 Low Income Nutrition and Diet Survey (LIDNS) of UK diets, published by the Food Standards Agency (FSA), reveals that low income population eat less fruit and vegetables (Nelson *et al.*, 2007).

[38] Both Dutch and EU policymakers have taken note of the need to promote healthy eating through catering. In 2005 the 'Overweight covenant' (a Dutch public-private partnership initiative) was signed by Veneca, the branch organisation for contract caterers. Through this programme caterers were encouraged to offer healthier food in work place canteens, as well as promote an overall healthier lifestyle, to pupils, students and employees.

Research within Europe and the Netherlands has focussed mainly on sustainable food procurement policy procurement, with the most well-known studies looking at sustainable food procurement for schools (Morgan & Sonnino, 2007; Sonnino, 2009; Sustain, 2005). The importance of catering services in shaping everyday sustainable food practices has not been looked at sufficiently. Canteens are important daily 'locales' of food consumption and caterers can be seen as 'orchestrators' of food consumption practices; they determine the content and taste of daily meals of millions of people. Not only do caterers influence daily eating habits, they also influence meanings and frames of reference of food. The catering study is discussed over two chapters. Both this chapter (Chapter 5) and the next (Chapter 6) are dedicated to investigating the opportunities and bottlenecks of sustainable development of food provisioning and consumption within work-place canteens. This should also reveal the wider relevance of the catering sector in shaping sustainable food practices within society. Chapter 5 focuses on the system of provisioning of work-place canteens, whilst Chapter 6 considers what is happening within the locale, the everyday social practice of eating in a canteen. It tackles the question of changing canteen food cultures and practices. It also considers the position of the canteen end-user in relation to (sustainable) canteen provisioning and providers.

In terms of the SPA framework this means we are looking from 'right to left' (Figure 5.1). In order to analyse the system of provisioning of work-place catering the following topics were researched: sustainable food procurement policy, catering contracts and criteria setting, food sourcing, canteen infrastructure and kitchens and the roles of different actors. The results of this study are presented in the following manner. First, a number of important limiting factors in sustainable food provisioning are discussed (Section 5.4). These are based on the findings from interviews and the focus group with caterers. Then a number of sustainable provisioning strategies based on examples of innovative initiatives are discussed (Section 5.5).

The research questions being posed in this chapter are:

- What are the major factors determining the success of sustainable canteen food provisioning?
- What kinds of strategies can be employed to realise sustainable canteen provisioning?

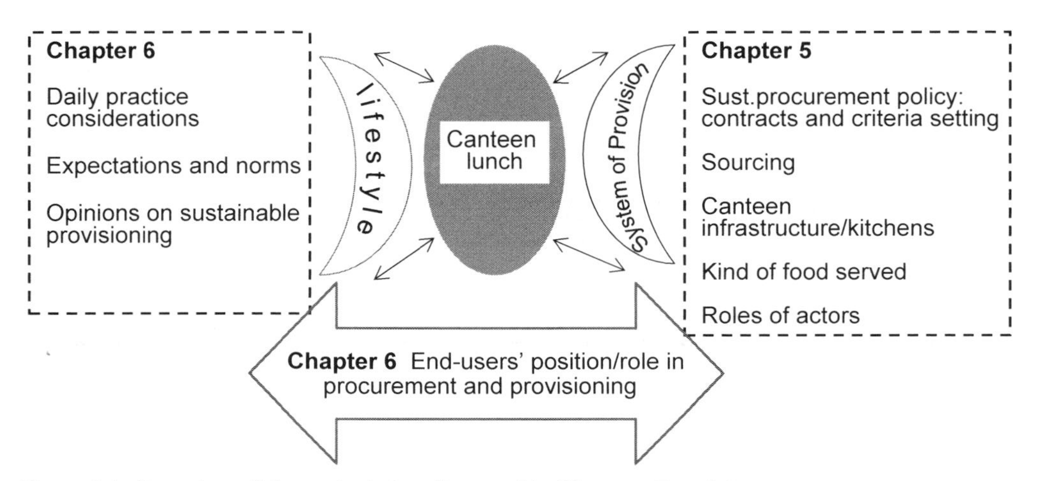


*Figure 5.1. Overview of the topics being discussed in Chapters 5 and 6.*

## 5.2 Methodologies as used in Chapters 5 and 6

### 5.2.1 Research focus

The focus of this study is on work-place catering, that is, the catering services for public and private organisations (business or company catering) and catering for institutions within Dutch higher education. Institutional catering (hospitals, prisons, elderly homes) and in-flight catering are left out of scope. Around half of all Dutch employees have access to canteen facilities, which amounts to approximately 2 million meals served in Dutch work-place canteens every day (BHeC, 2005). In addition, most students in higher education and vocational training have access to some sort of canteen. In cases were catering services are not available, especially in smaller organisations, lunch may be provided via other types of external catering delivery services. These services are also left out of scope.

There are three types of actors directly involved in deciding what kind of food is provided in canteens and how this it is organised. These are the caterer, the employer of the caterer, which is the contract-lending party in cases where catering is hired in, and the end-user (which in this case are students and/or employees). The primary focus is on the caterers and the end-users. Initially, the choice to concentrate mainly on the caterers instead of their employers/contract-lenders was motivated by the fact the objective of this study is to concentrate on what is happening within the practice of canteen food consumption and its context. This context is most directly influenced by the caterer, who organises, supplies, prepares and promotes the food; thus having a substantial role to play in the 'framing' the canteen food provisioning. Information on the role of the contract-lenders is included though mostly gained from respondents who work in catering.

### 5.2.2 Methodology

Chapter 5 covers an analysis of the systems of provisioning. This is based on literature research, interviews and the focus group with caterers. Chapter 6 covers the everyday routine of canteen food consumption practice and what kinds of considerations are made by end-users, also on the topic of sustainability. It is for the most part based on the focus group with canteen end-users (employees and students of a Dutch university). The discussion on the issue of how sustainable food consumption (Section 6.6) should be encouraged in on the canteen floor is based on both the focus group with end-users and the focus group with caterers (Table 5.1 question c).

*Literature research and interviews*

Literature research was carried out to understand the Dutch catering sector and its major features and trends. First, literature was sought on the developments in sustainable procurement policy and criteria setting for public and private organisations as encouraged and supported by the government. A literature study was also dedicated to finding various innovative initiatives and caterers active in sustainable catering in the Netherlands and Europe. A number of these innovative examples were then selected and visited, most were accompanied by interviews.

*Table 5.1. Three main topics discussed during the focus group sessions.*

| Questions posed to caterers | Questions posed to end-users |
|---|---|
| a. What do you do to realise sustainable provisioning? What opportunities and bottlenecks do you perceive in relation to sustainable provisioning? | a. What does your lunch routine look like? What do you think about the level of sustainable food provisioning in your canteen? |
| b. What role do you see for yourself as a caterer in realising sustainable provisioning? | b. Do you believe you have a role to play in realising more sustainable provisioning? |
| c. How do you think consumers should be encouraged to make more sustainable food choices in the canteen? (+ discussion in small groups) | c. How do you think providers should approach consumers in order to encourage sustainable food choice within the canteen? (+ discussion in small groups) |

Interviews were semi-structured and held over the period 2006-2011. The following actors were interviewed:

- catering managers;
- cooks (in charge of catering);
- NGO's involved in sustainable catering initiatives;
- facility management consultants;
- researchers.

A number of canteens were visited mostly in combination with interviews with the caterer in order to investigate:

- the canteen system of provisioning: food supply and kitchen infrastructure;
- the canteen locale: the kind of food served, appearance and set-up of the canteen (the canteen 'concept').

The topics covered during these interviews (Appendix 2) were (canteen system of provisioning, Figure 5.1):

- Sustainable procurement policy, contracts and criteria setting. What kind of sustainable procurement policy was used by the employer/contract-lender and/or the caterer. How sustainable provisioning of canteens is/can be realised (bottlenecks and opportunities).
- Sourcing of food. What kind of sustainable foods (e.g. organic, Fair Trade, etc.) are available in the canteen. What kind of suppliers were used and why.
- Kitchen and canteen infrastructure.
- The kind of food served in the canteen, i.e. meals and range of foods on offer in the canteen.
- What role different actors play in this process.
- Perceived bottlenecks and opportunities for furthering sustainable food provisioning.

The research focuses on sustainability in relation to the food served in canteens, and not on sustainability in relation to the food preparation techniques or appliances (e.g. energy and water saving). Sustainable canteen food provisioning refers to the provisioning of foods which are demonstrably beneficial to either the environment and ecological systems, and/or beneficial to animal welfare and/or the livelihood of local communities and/or primary producers in third developing countries. Most caterers interviewed and those who participated in the focus groups use organic and Fair Trade labelled products in their sustainable provisioning strategies, as well as seasonal and local sourcing. In that respect, the definition of sustainable provisioning employed here is similar to that set out in the guidelines on sustainable procurement for catering (VROM, 2010).

### *Focus groups*

In 2007 two focus groups were held: one focus group with caterers[39] and one with end-users. Using the focus group methodology allows one to acquire information about how a group of people think or feel about a particular topic, and gives insights why certain views are held (Bloor *et al.*, 2001; Morgan, 1988). Group discussion can stimulate respondents to give more information on their opinions providing insight into different rationales and arguments. It is not a reliable way of gathering information on individuals or collecting representative data. Respondents may influence one another and group dynamics may have the effect of encouraging respondents to speak just as much as it may prevent some to give their opinions. A script and focus group facilitator were used for each focus group.

Participants of the first focus group consisted of seven participants. These included caterers from a Dutch University (in-house catering services), a Dutch vocational college (in-house catering services)[40], a Dutch bank (in-house catering services), a Dutch municipality (external catering services using contract-caterer) and a catering manager from a major Dutch contract-caterer. This gave a good spread of caterers from different kinds of organisations and catering services. The participants were selected because they were known to already be involved in sustainable catering.[41] Participants were invited via a formal letter.

The second focus group consisted of a group of eleven canteen end-users. The participants were employees and students from a Dutch University. These participants were recruited by announcing the focus group session on the university bulletin board. Table 5.1 shows which topics were discussed during the focus groups. Both caterers and end-users were asked a number of questions, which were then discussed within the group.

[39] Other than the ones interviewed.

[40] This is actually an in-house caterer that is part of another company.

[41] Most of them had signed the Dutch public-private partnership covenant for the Market development of Organic Agriculture (discussed later on in the chapter).

## 5.3 The systems of provisioning

### *5.3.1 Types of catering services*

There are various different ways in which organisations may arrange their catering. Some organisations have their own catering services, referred to as in-house catering, these are part of the internal facility services of an organisation (Figure 5.2). In-house catering services may be organised differently in different organisations, but generally it is the caterer who makes decisions about which suppliers to use and what type of foods (and recipes) will be served. Some organisations hire a caterer to operate food services within their canteen facilities (contract-catering). These caterers can cook on site if the canteen has kitchen facilities for cooking. There are many small, locally and regionally operating caterers, but a large part of the market is dominated by the larger contract caterers who operate on an international scale. Lastly, there are organisations with more than one canteen or restaurant/café, and here various external caterers/food providers may be active and/or a combination of in-house catering and hired catering.

Over the last three decades there has been a general shift from subsidised, self-organised catering (in-house) to hired or contract catering, that is, outsourcing catering services to more commercial catering companies, a trend which is seen in other European countries (examples are Scandinavia and the UK) (Morgan & Sonnino, 2007; Post *et al.*, 2008). For instance, in the Netherlands the number of canteens contracted to external caterers has increased steadily between 1994-2008 (BHeC, 2004). Within Dutch higher education figures show that 69-93%[42] of the main catering services are outsourced (Daaen & Van der Meer, 2010) and catering within

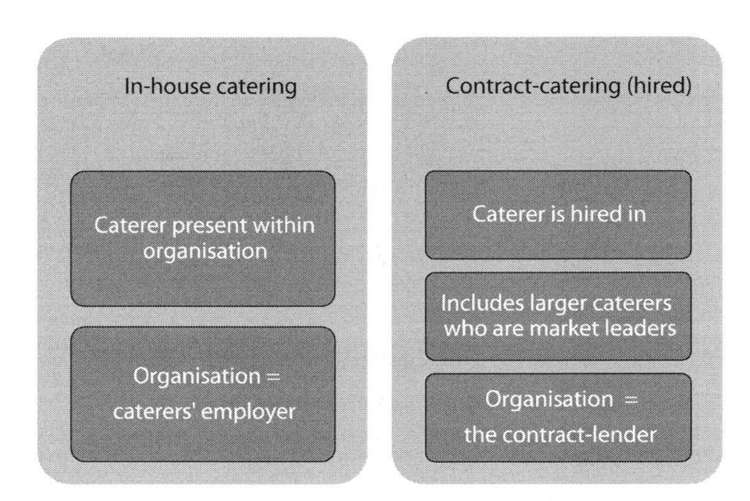


*Figure 5.2. Generalised representation explaining the difference between in-house catering and hired contract catering providing catering services within an organisation.*

[42] Colleges/polytechnics and universities (ROC=71%, HBO=93%, WO=69%).

educational institutions is less subsidised than catering services offered within government authorities, businesses and health care (Damen, 2010a, 2010b). Whereas in the past canteens where often subsidised, today canteens are largely financially independent and the caterer held accountable for the finance of their catering services. However, caterers often still receive some form of support, for example, through discounts on rent (interview one caterer). Also, especially in the case of smaller caterers, labour costs may be charged to the contract-lending party or the latter takes on some of the burden of these costs.

A small number of large contract caterers dominate the Dutch business catering sector (Sodexho, Compass Group, Albron and Elior, with a total market share of 85% in 2004) (Factcard, 2004). In 2009, contract-catering (or hired catering) had a share of 33% of the European social foodservices market[43] value (FERCO, 2009) and in most European countries business catering is by far the most important catering activity for contract-caterers. Two important drivers for hiring caterers from outside are the need to cut overhead costs (spending lump-sum money more efficiently) and the increasing regulatory complexity involved in running a canteen. The latter refers to hygiene regulations, quality control and public procurement regulations. For example, in 2004 caterers had to adhere to 100 binding regulatory rules and more than 75 administrative responsibilities created by more than 15 different organisations (BHeC, 2003). Also, food procurement has become more formalised with compulsory competitive tendering procedures in the public sector and more recently European regulations on public tendering.

### 5.3.2 Actors shaping canteen consumption practices

The three major actors who 'shape' the canteen food consumption practice are the caterer, the caterers' employer (in the case of in-house catering) or contract-lender (in the case of hired catering) and the end-user. In general, the caterer organises the daily food service, determines what foods will be served and how they will taste. The employing organisation (in the case of in-house catering) or contract-lender (in the case of hired catering) is mostly in charge of setting a budget for catering and on deciding whether catering should be outsourced or not. In the case of in-house catering, the caterer is often the one who determines and applies the more or less formal food procurement criteria, that is, what type of food products to source and where to source them from. The organisation may or may not have a formalised food procurement policy which effects these practices. In the case where caterers are hired from outside such procurement criteria do need to be formalised in order to come to a contract between the two parties. This is the task of the facility management within an organisation. Often such criteria are related to price, sustainability and/or health criteria, the kinds of foods and meals served, and the range of choice in the food served in the canteen. Where caterers (should) source their food from is often not specified within contractual agreements between the caterer and the contract-lender. In addition, caterers themselves can have their own set of criteria and policies with respect to the procurement of ingredients, food products, food suppliers and wholesalers, etc.

[43] Social food services refer to catering services for education, health care and elderly care and public/private organizations.

In all this the caterer is the 'food expert' who largely determines the type of food served the atmosphere and 'culinary experience' within the canteen. The caterer interacts with customers and has the kind skills and knowledge about food and hospitality which make him the primary actor in shaping food consumption practices within the locale of the canteen (Figure 5.3). The interaction between the contract-lender/employer and the end-user on the subject of food services is often limited. End-users are not often involved in the actual process of the organisations' food procurement processes and policies, although some organisations may have user-committees which may be invited to be involved in the tendering procedures and/or consulted as a kind of consumer panel. The interactions between the end-user and the caterer takes place mainly within the canteen locale, the consumption junction, but not usually in more formal related settings where end-users are involved in making choices about food provisioning and procurement. Many caterers do engage in customer satisfaction surveys besides the daily personal interaction with customers.

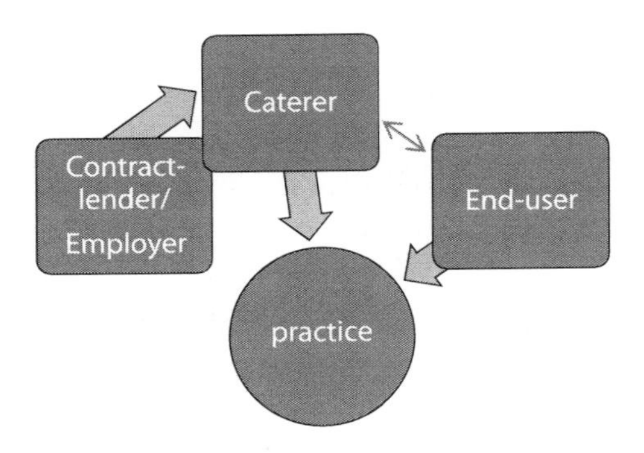


*Figure 5.3. Actors shaping canteen consumption practice and general representation of the interactions between actors.*

## 5.4 Factors influencing the success of sustainable food provisioning

From the literature study, the interviews and the focus group with caterers a number of reoccurring themes arose which will be discussed in this section. These factors are thought to play a key role especially in influencing the process and success of introducing sustainable foods (like for instance organic and Fair Trade food products) in the canteen.

### *5.4.1 Size and the role of contracts*

Large contract-caterers operating may be less flexible than the smaller caterer to change their procurement and sourcing practices, which can present a barrier to realising more sustainable food provisioning. Many of the large catering companies (contract caterers) have professional buying departments which make deals with suppliers (wholesalers, producers) within and outside of the Netherlands, often receiving, and depending on, discounts from suppliers. Thus not only are there contractual agreements between contract lender and caterer, but there are also contracts and discount incentives with suppliers (Figure 5.4). This structure can make it more complicated for catering managers operating on the canteen floor to make (day-to-day) changes within food provisioning. They do not have the freedom to try out various different suppliers (see also GM&FNE, 2007) and make daily 'on-site' changes, i.e. changes in recipes, make special offers and/or introduce new products.

Especially the larger contract-caterers work with more standardised work processes and recipes, and contracts are set in such a way as to limit risks and costs. Hired (contract) catering can provide more control over spending on facility services. Some contract forms even allow for all risks to be borne by the caterer. The content of the services are described in great detail in one official document which forms the basis for the contractual agreement (the terms of agreement, in Dutch 'Programma van eisen'). This includes for instance a detailed list of the range of food products which are to be included in the lunch buffet, e.g. 3 kinds of sandwich toppings, 3-4 kinds of different dairy drinks, etc. Agreements are made on the prices and content of all of the different services provided, for instance, the food served during lunch hours in the canteen, what kind of lunch can be ordered for groups/meetings, etc. The system is arranged in such a way as to keep what is on offer in the canteen constant, as the goal is to have the same food products on offer to consumers every day. Food wastage is carefully monitored in order to realise cost-efficiency. If sustainability is to be considered within this system it needs to be taken on board from the early stages of the whole procurement and contract process and built in solidly within the details of the terms of reference.

There are different types of contracts between the caterer and the contract-lender. With an open book contract all costs (risks) and profits are for the contract-lender and the contract-lender decides on which products will be provided and the margin on these products. The caterer is hired and paid a periodically calculated management fee. In the case of the provider contract operations are either fully or partially in the hands of the caterer and thus both risk and profits are for the caterer. Over the last ten years there seems to be trend towards the use of this type of contract, since its benefit lies in that caterers take full risk (VENECA, 2007) and it offers organisations (contract-lenders) more security and stability.[44] All details on exploitation costs and returns must

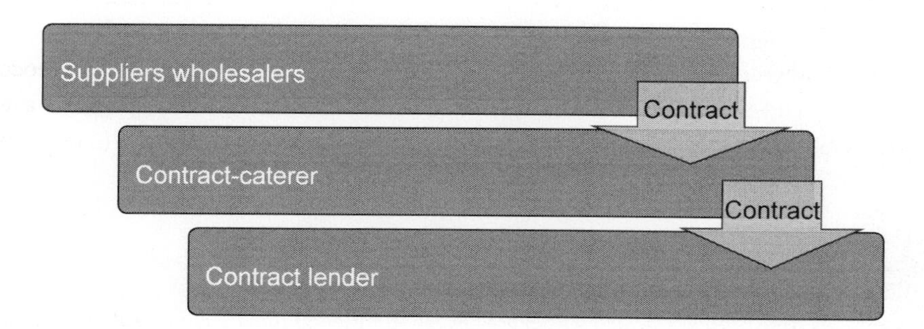


*Figure 5.4. Generalised representation of contracts between supplying parties, caterers and contract lenders in the case of contract-catering.*

[44] For example, changes in labour costs, changes in food provision costs and changes in rules and regulations do not have to be dealt with any longer.

be worked out prior to the signing of the contract. On the one hand this gives the caterer more flexibility to deal with changes in the supply of products, and for instance, temporarily switch to another product (VROM, 2010). On the other hand, it is important for the caterer that costs remain the same during the duration of the contract, since any discrepancies will fall outside the contract. This functions as an added incentive to keep costs low.[45]

### *5.4.2 Canteens and kitchens*

In many canteens there is a partial de-coupling of food preparation. Many canteens have an infrastructure that it is unsuitable for preparing and cooking fresh foods (or 'cooking from scratch'). They do not have a proper kitchen and or skilled cooking staff. Many caterers have adopted standardised meal systems, often with the use of processed, easy-to-prepare foods.[46] Their added benefit is that they make it easier to keep to health and safety regulations. These standards state that leftover perishable foods kept outside of the fridge for a number of hours during lunch hours should be disposed of (NEN-ISO 9001:2000, Hygiene Code for Contract Catering 2004, 2007).[47] This creates a situation where there is an added incentive to use small-volume (mono) packaging. Some caterers monitor the amount of food they waste and this can be used as a performance indicator.

The fact that many canteens are without a properly functioning kitchen (i.e. a kitchen for cooking) limits the range of possibilities caterers have in food preparation and therefore also limits the different options available for creating more sustainable provisioning (He & Mikkelsen, 2009).[48] During interviews, various contract caterers spoke about their dependency on canteen infrastructure available to them. They mentioned the size of the canteen and buffet, the kitchen infrastructure (kitchen or heat-up facilities) and the presence/absence of waste disposal and/re recycling facilities. Without a kitchen suitable for cooking, preparing fresh warm foods (e.g. soups, meals) is near to impossible. Caterers simply have less options to experiment with different recipes, limiting their ability to be creative and flexible. A kitchen-less canteen will be more dependent on ready-made foods (CGE&Y, 2002) than one which can produce its own meals (Figure 5.5). This canteen will therefore be more dependent on the availability of sustainable alternatives within the supply chain, i.e. the ready-made foods/food components made available be catering suppliers and wholesalers (see also Post *et al.*, 2008). Overall, this lack in proper infrastructure, food preparation and cooking skills mean such canteens are simply less able to organise their own sustainable provisioning, and to add value to this provisioning.

[45] There is also a contract form where both losses and profits are shared by the caterer and the contract-lender, and a more commercial form where the caterer is operating in an independent manner, paying rent to the contract-lender. The latter may specify terms on price and the kind of food served.

[46] Processed foods have been implicated with a greater environmental impact compared with unprocessed foods (Engstrom & Carlsson-Kanyama, 2004).

[47] Such kinds of hygiene regulations are said to be part of the cause of the large amount of food wastage in the catering industry. There are discussions on how to deal with and/or change such regulations.

[48] He and Mikkelsen (2009) report on their project about organic school food provisioning that the main challenges for an increased consumption of organic food in schools are related to lack of infrastructure in the schools such as kitchens and dining halls.

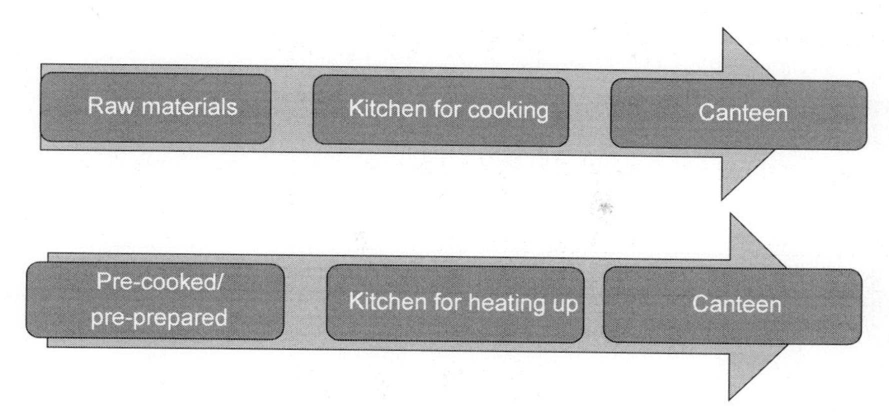


*Figure 5.5. Generalised representation of two canteen infrastructure typologies.*

### 5.4.3 The working relationship between caterer and contract-lender

The two parties, contract-lender/employer and caterer, must work together to achieve sustainable provisioning, and thus the nature of this working relationship is important. Especially in the case of hired catering, where the actors are part of different organisations, this process can be seen to entail a negotiation of where responsibilities lie and 'who will pay' for realising more sustainability. One caterer commented: 'In the end the contract-lender pays. That can sometimes be a handicap, because you might have your own ideas about how to improve the catering, but the contract-lender must agree with these ideas'. Another caterer: 'If we want to further develop sustainable canteen food provisioning we need the support of our contract-lender. You have to do it together'. Some contract-caterers indicated that it is difficult to make their own sustainability policy when they are dependent on the contract-lenders ideas and wishes in relation to sustainability. They felt that on this issue they are limited in their entrepreneurial freedom. They can only function as middlemen and cannot 'be much more sustainable' than contract-lenders allow. Some caterers spoke of the lack of investment in food services made by organisations.

One participant spoke about the 'attitude' organisations can have towards canteen food services. According to him work-place catering is an inherently functional service and procurement officers and facility managers tend to be concentrated on saving costs and efficiency (see also Heerkens, 2009: p. 52): 'Generally the people caterers work with are buyers and facility managers. These tend to be technocratic and reason in terms of price and functionality, and less in terms of food experience and atmosphere. That means there is less space for changes to the canteen'. Another caterer commented: 'Catering is seen in a very process-oriented way and not enough in terms of business opportunities ... like in the restaurant sector'. According to this participating caterer, business catering is still not commercial enough. Any cost savings made in the procurement phase usually comes to the good of the end-user (i.e. low canteen prices) instead of the caterer which can work as a disincentive to entrepreneurial creativity and value adding. This phenomenon is also mentioned by one facility management expert, who indicated that caterers complain about the lack of return due to relatively low margins hampering their ability to really offer better quality/more sustainable products. They point to the low willingness of contract-lenders to invest in catering and end-users' low willingness to pay for canteen food, for instance because of the perception that canteen food should be cheap.

According a facility management expert, improving catering services centres requires an improvement of the working relationship between caterers and contract-lenders, and a different orientation toward the target-group/the end-user (Damen, 2010a; Damen & Van der Meer, 2010). Sometimes organisations which have outsourced catering are not sufficiently 'in touch' with its catering services (Damen, 2010b). Contract-lenders/employers need to maintain an active role, for example by providing the necessary stimulus and/or room to allow caterers to innovate and develop in improving sustainable provisioning (see also Damen & Van der Meer, 2010), and/or to make sure that caterers are hired who are able to deliver on these aspects. Damen and Van der Meer (2010) indicate that the employer/contract-lender needs to see catering as an investment which can generate return, instead of a cost. Part of this means thinking in terms of service quality and 'customer service'. This in turn means considering the target group, and a good fit between food provisioning and the end-user (Chapter 6).

This is where the caterer plays the most important role. It is the caterer who brings the knowledge of food service to the table. The caterer is the one who designs the catering concept and who has ideas about 'creating food experience'. Caterers provide food for what can be called 'captive consumers'. As one researcher put it: 'Within catering there are more links operating within the chain than in retail, and they all have an influence on making decisions about the food provided in the canteen. However, especially the catering manager has a key role to play ...' (interview).

### *5.4.4 Canteen food culture: habits, tastes, norms and expectations*

During the focus group participating caterers were asked how they perceived their role as a 'change agent', i.e. having an active role in 'greening' canteen food consumption.[49] Concerning sustainable food provisioning, do they see themselves as 'followers', i.e. responding to the demand for sustainable products, or as 'innovators', i.e. actively shaping the demand for sustainable products? The overall response was that in theory caterers should be leaders in setting new food trends, and that there was great potential for the role of catering in sustainable development. However, in practice they are dependent on what clients and consumers want. One caterer participating in the focus group explained that, unlike retailers and restaurant holders, caterers cannot 'choose' their consumers, making it relatively more difficult for them to introduce sustainable provisioning. 'The caterer is bound to a certain target group' (where the caterer happens to operate). One caterer described contract-catering a 'passer on of products', i.e. a middleman, simply answering to the needs of their customer. Some caterers were of the opinion that the drive for sustainable provisioning is mainly demand-driven. That is, not just the demand from contract-lenders/employers and from end-users, but also the demand for sustainable food products in society in general. For example, on the subject of organic food caterers spoke about the image of organic products as something which is shaped external to the canteen, dependant on the trends for organic products within

[49] The questions posed in the focus group was: 'What role do you see for yourself as a caterer in realising sustainable provisioning?').

society. According to this reasoning organic products first need to become more mainstream before they can offer it in their canteens.

The lunch and canteen food culture in the Netherlands generally, and the food culture within an organisation were mentioned as factors that play a role in whether or not introducing more sustainable foods will be successful. Some caterers spoke about the general attitude toward canteen food and Dutch lunch culture. The Dutch spend on average 2-3 euro's on their lunch, whereas the Belgian, for example, spend 5-6 euro's (interview). Some caterers feared that this makes the introduction of more expensive sustainable alternatives more problematic.[50] Also various caterers spoke about their reluctance to change menus which their customers are used to, as changes can lead to disappointment, protest and/or a loss of clientele. One caterer who participated in the focus group spoke about how end-users (taste) expectation and standards play a role in how they react to certain changes. He suggested as an example that end-users who are used to freshly made soup will not like to go back to pre-packaged soup, they will perceive the taste difference. Whereas people who are used-to packaged soup for their lunch will be happy getting pre-packed soup. In other words, people get used to certain tastes which influence the way they react to certain changes in provisioning.

Canteen food culture does not primarily consist of the practices, standards and expectations of the end-users. Canteen food providers also contribute to, and uphold, a certain food culture. On the one hand, there are the practices and attitudes facility managers and contract makers, which have been shortly discussed in Section 5.4.3. On the other hand, there are practices and attitudes of the caterers themselves. Caterers have conceptions about what foods should be served in work-place canteens, what consumers want and how they will react to change. A good example was given by one catering expert concerning party catering. There are some foods which are considered typical 'party' foods and are expected to be served (for instance strawberries, even in the winter, and shell fish). Since these are the standards some caterers might not be inclined to substitute them, some might feel that changing such standards is equal to 'bad service'.

One of the caterers interviewed commented on how realising more seasonality in canteen food provisioning sometimes clashes with the culture of standardisation. Many canteens operate via a standardised ordering and food preparation scheme, and what is offered in the canteen on a daily/weekly basis is also kept fairly constant throughout the year. This is a convenient and efficient way of operating, but it also mirrors a certain approach to the consumer. The idea is that generally people do not want to veer from the lunch content they are used to, and caterers keep to the routines and habits of their customers, serving the kind of food which (they perceive) is expected of them. In this way caterers feel that they are catering for the wishes of their customers. Changing such a catering system and ditto food culture requires effort and time. In such cases introducing more seasonal sourcing means consumers have to get used to more flexible and creative provisioning.

Thus, there are certain norms related to catering services, both the norms of consumers and providers, which might need to be questioned or revised if provisioning is to become more

[50] One respondent knew of a canteen where end-users organised a small protest after the caterer had made the deep fried snacks ten cents more expensive in order to encourage more healthy eating patterns. In the past the government has encouraged caterers to increase the price of fattening foods in the spirit of healthy eating policy (Overweight Covenant, 2005).

sustainable. Such norms may relate to various aspects of the catering operation, from sourcing, to menus and recipes, but also food preparation and storage methods.

## 5.5 Sustainable food provisioning strategies

Whereas the previous section looked at some of the most pressing limiting issues involved in sustainable canteen provisioning, this section concentrates on a range of different strategies to realise more sustainable canteen provisioning, providing examples of innovative initiatives.

### *5.5.1 Sustainable procurement policy; criteria setting for sustainable contracts*

In various European countries, like the Netherlands, Scandinavia, UK and Italy, sustainable public food procurement has made an appearance on the agendas of policymakers and government officials. In the Netherlands, it started in 2004 and 2005 with the Dutch parliament accepting the motion by Koopmans and De Krom to apply governmental sustainability criteria to all aspects of public procurement. In this way the public sector hoped to stimulate the sustainable economy and provide an example to private sector procurement. Since then public procurement criteria have been developed for all areas of procurement, including catering. Facility managers and caterers can use the guidelines on criteria to help them select particular foods and suppliers, but also in selecting catering services when catering is hired in from outside the organisation.

Both national and EU documents on criteria offer guidance on the kinds of social and environmental criteria which could be used in sustainable food procurement. In 2006 the Ministry of Environment (VROM) and the Former Ministry of Agriculture (LNV) (now the Ministry of Economic Affairs EZ) set up a working group with various public and private stakeholders[51] to create a set of guidelines on sustainable procurement for catering (VROM, 2010). They are revised almost each year and are primarily applicable to the public sector[52], and can be applied to cases where catering is hired and in-house catering. Sustainability criteria can be taken up on a wide range of different issues (Table 5.2).

Large organisations and (semi)governmental organisations often hire a caterer via a tendering procedure. This procedure can be public, whereby any caterer can register and then make an offer. In non-public tendering procedures ('onderhandse aanbesteding') two or three caterers are invited to make an offer. Via a contract, agreements are made on the prices and content of the food service (i.e. what is served during lunch hours, what lunch can be ordered for groups/meetings/ parties, etc.), but it can also cover other topics, such as how the caterer will engage in promotion and sustainability. Above a certain value the tendering procedure falls under the EU procurement

[51] The stakeholders included: a municipality, a Water board, a consultancy company, Veneca, a wholesaler and two Dutch NGO's (Natuur en Milieufederatie, Stichting Milieukeur).

[52] Food procurement by the public sector is required to be 40% organic or other 'sustainable foods' (e.g. MSC fish, SMK (Dutch environmental label) or regional speciality products (Erkend Streeekproduct is the Dutch label for geographically characteristic produce). Other topics include waste and cleaning methods, social criteria, as well as a guidelines as to how to incorporate criteria within the entire process of procurement (VROM, 2010).

*Table 5.2. Aspects covered by the sustainable procurement criteria for catering (VROM, 2010).*

| **Production drinks and food** |
| --- |
| Health |
| Packaging and waste |
| Energy and water use |
| Cleaning |
| Working conditions |
| Workers participation |
| Delivery |

regulations, and the tender is open to any caterer within the EU (VENECA, 2007).[53] The EU encourages the inclusion of sustainability criteria and besides the main directive on procurement (EC, 2004a) interpretative communications exist which give directions on integrating social and environmental criteria (e.g. COM (2001) 566 and Com (2001) 274) (EC, 2011; Morgan & Morley, 2002: p. 26).[54] This means that buyers can select not only for the most competitive supplier or most competitive tender, but also according to certain environmental and social standards (Eurocities, 2005; VROM, 2010).

Realising sustainable food procurement requires knowledge about contract setting and the skill to realise ones sustainability goals/wishes. Various studies show the importance of skilled professionals playing a key role in implementing sustainable procurement criteria (Adrichem, 2009; Eurocities, 2005; He & Mikkelsen, 2009; Morgan & Morley, 2002). An added challenge which many organisations face is to realise sustainable food provisioning without substantially increasing costs. This may require a process of learning and/or hiring/involving experienced procurement officers, managers or food professionals (see also studies by Adrichem (2009) and Heerkens (2009)).

### *Criteria on local provisioning*

'Near-souring', which is cutting down on 'food miles' by buying foods from more local/regional suppliers, appears to be a popular way to engage in sustainable food procurement. Various examples exist of local government and caterers who include regional and local aspects in the foods they buy. For instance, local government programmes in France, UK, Italy and Scandinavia link sustainable development to localisation (Mikkelsen *et al.*, 2002; Morgan & Sonnino, 2007). To what extent near sourcing is compatible with EU regulation remains a grey area. One caterer who works with local and regional suppliers/producers indicated that categorising such activities as sustainable

[53] Depending on the type of public organisation this value ranges between 137,000 and 422,000 Euro. Under the EU threshold public procurement follows regulations which are different for different municipalities.

[54] COM (2009) 215 looks at the integration of fair trade standards (EC, 2009).

'is a sensitive issue because of the EU'. Another caterer indicated that with the coming of the European procurement regulations he could only engage in near-sourcing for smaller orders, since the larger orders need to follow EU regulation. Regulations do state that criteria applied should be transparent and non-discriminating (EC, 2004a; SenterNovem, 2010), in other words, that public procurement must not be used explicitly to support domestic producers. Thus, generally speaking EU procurement rules do not allow specifications to 'local' food in public catering contracts (Morgan, 2008; VROM, 2010).

However, public authorities have found ways to include near-sourcing within contracts. Conflicts with regulations can be avoided by designing various smaller contracts instead of creating one large contract. Also, any policy goal (environmental or social) which an authority is trying to support via its procurement must be explained and phrased in terms which demonstrate the relevance of a procurement choice (e.g. Fair Trade products) (Eurocities, 2005, p. 20). Public organisations in Italy and France have designed contracts with specifications on food quality arguments citing such aspects as freshness and seasonality, as well as the wish to serve organic food, as reasons for preferring local food suppliers (Morgan & Morley, 2002; Morgan & Sonnino, 2007).

### *Contextualising food procurement*

In fact, food procurement is well suited to incorporate a range of different and/or interlinked sustainability goals (economic, social and environmental). The public procurement for Italian schools is a good example of this 'contextualisation' of sustainable catering (Morgan & Sonnino, 2007). In Italy sustainable provisioning is connected to preserving Mediterranean food culture, Italian food traditions and organic farming. Italian public food procurement policy is clearly opposed to highly processed, mass produced food (p. 21). There is a multifunctional view on school catering, namely to realise healthy diets for children and educate them on Italian food culture. In this way sustainable catering is imbedded in the kind of food culture which is recognisable for Italians. In addition, food procurement policy is carried by a structure of different stakeholders, thus creating a broad support for and (social) monitoring system of canteen meals. Parents, schools and the municipality work together within the food procurement system of school canteens.

What can be learned from the 'Italian model' is that sustainable food procurement policy can benefit from connecting procurement to shared food values and/or other sustainable development goals, recognisable to the public. '... depending on the nature of the provisioning, it can address social justice, human health, economic development and environmental goals, the main domains of sustainable development' (Morgan & Sonnino, 2007: p. 4). Examples of (public) sustainable food procurement programmes with such ambitions, such as the Dutch Proeftuin Amsterdam (Experimental Garden of Amsterdam), the Danish Dogme 2000 programme and the Scandinavia IPOPY programme, also show how sustainable food procurement can be linked to or framed in terms of a wider range of issues like food education, social learning, creating job opportunities in catering and dietary health and overall food quality (Morgan, 2008). Even issues such as traffic congestion and urban food production can be included. This may help create a broader platform of support from different parties. In the Netherlands the Biobites project is an example where two objectives are combined; healthy canteen food for youngsters and market stimulation for organic farmer in the region. This project is the result of a public-private partnership between the provincial authority of North-Holland, the wholesaler New Organic World and Cormet, to supply organic lunch (soup, sandwiches, milk and fruit) to Dutch secondary schools and vocational training schools (14-21 years old). The project aimed to improve school catering in 80 schools in the province of North-Holland using produce sourced from organic farmers in the region. Biobites was the first project of its kind in the Netherlands; introducing organic food in a Dutch secondary school whilst marketing it is 'cool' to appeal to teenagers.

### 5.5.2 Sourcing via wholesale

The availability of sustainable products within wholesale is an important factor in enabling sustainable canteen provisioning considering that many caterers source their sustainable alternatives (organic food, Fair Trade labelled foods, etc.) from there. Whilst some caterers prefer to source from different smaller-scale suppliers (farmers, producers), looking to get certain types of produce with certain quality or origin, many caterers prefer using a wholesaler as most or all canteen supplies can be sourced within one order. Buying products from as few suppliers as possible saves time and is convenient. For instance, it is much more convenient to have all the supplies come at the same time each day, with the same lorry, with just one bill to process. Caterers can get discounts with orders of a certain size. Large caterers, who cater for many canteens often have long-term contracts (5-6 years) with large organisations (contract-lenders), will receive a greater discount by setting one large contract with a wholesaler. In addition, one caterer mentioned that he finds it easier to control and guarantee food safety with just one main supplier.

There are a range of wholesalers who specialise in supplying organic food and other certified sustainable alternatives (Benschop, 2008) and amongst the larger mainstream wholesalers the availability of organic and Fair Trade labelled (amongst others) is growing. Two of the largest food wholesalers have set up special sustainable food concepts for their sustainable assortment.[55] This reflects the general demand within the sector for food products and produce which have an added value in terms various sustainability or quality related attributes. Both wholesalers are working on improving their range of product alternatives, so that caterers will be able to access a wide range of sustainable alternatives (e.g. MSC fish, Fair Trade coffee, organic vegetables, etc.) but also developing product lines which promise fresher produce directly from the farmer. The last few years large wholesalers have followed the trends in the market for more authentic and fresh food produce, for example, offering bread made with more artisanal recipes.

Issues most often mentioned by respondents in relation to the provisioning of sustainable alternatives are price, availability (supply and quality) and demand (amongst end-users and contract-lenders/employers). The extent to which each of these issues play a role may differ for different kinds of foods and canteen systems. For instance, replacing bread, dairy products and coffee with their organic/Fair Trade alternatives has proven relatively easy over the years (GM&FNE, 2007) compared to convenience products such as ready/pre-prepared meals and meal-components (Ernst, 2007; GM&FNE, 2007). Canteens without a proper infrastructure for

[55] Sligro and DeliXL, two large Dutch wholesalers, are working to improving their product range and information provisioning on products. Via their websites orders can be made and detailed information on all the products can be found.

cooking, that is with a fully functional kitchen and skilled cooking staff, and/or a preference for convenience foods must find sustainable alternatives for pre-prepared soups, meals and sauces. However, these might not always be available in the taste and quality which is whished.

### *5.5.3 Near-sourcing: means to an end*

Caterers may take up near-sourcing as a part of their sustainable procurement activities. Different definitions of near-sourcing are used. Some speak of local or regional sourcing in terms of 'saving on food miles', while others refer to (re)localising and regionalising their supply. Sometimes it is used in reference to a particular product from a particular region (e.g. Wadden Goud, Erkend Streekproduct). Near-sourcing appears to be somewhat of a trend within the hotel, restaurant and catering sector. Even Dutch mainstream wholesaler Sligro (Eerlijk & Heerlijk) has taken up some regional speciality products (Streekproducten) into their product range.

Various respondent caterers have indicated that sourcing (some of) their produce from smaller, often more local/regional suppliers and/or producers is part of their strategy to realise sustainable provisioning. Caterers give different reasons as to why they engage in near-sourcing. For some saving on food miles and $CO_2$ emissions seems a major consideration. Others want to support local/ national businesses who produce in a certain way (e.g. more artisan, healthy and/or sustainable), or speak about the reconnection of consumers to the rural area (Box 5.1). For entrepreneurs who see the rural areas around them becoming degraded and losing business near-by sourcing is a way to invest in the future of where they live. There are many initiatives in the Netherlands on (re) localising and regionalising food networks, many centring on this 'social' sustainable development principle (e.g. Van Flevolandse Bodem).

In many cases, however, near sourcing is not so much described as a goal in itself, but is seen in a more functional light; to realise more transparency, better taste, more seasonality, culinary quality and authenticity, thus adding value to the service (something which is almost always mentioned). Various caterers described near-sourcing as a way to be able to afford organic food (e.g. catering in Ter Reede elderly home) (Ernst, 2007). Money can be saved by cutting out 'the middleman' (e.g. some wholesalers, or contract caterers) by sourcing from producers, producer cooperations, or local wholesale businesses with the logistics to source regionally. In the case of the catering for elderly home Ter Reede (kitchen, in-house catering) in the Netherlands the goal is to maintain 100% organic food supply, whilst being cost neutral. According to the experiences of this caterer, this can be realised through regional/near sourcing and looking at cutting down on food wastage through optimal use of all the food (e.g. using all of the broccoli, not just the rosettes, requires being creative).

One of the major problems caterers face in relation to near-sourcing is the availability of local/regional produce in adequate quantities and satisfactory. In many parts of the country the logistics of getting locally produced products to canteens is underdeveloped. Other issues are the inadequate volumes, the right timing and/or consistence in supply from smaller, regionally and locally operating suppliers. In addition, it requires more effort on the part of the caterer to deal with various suppliers compared to making one order from a wholesaler. One caterer mentioned that many primary producers fail to invest in proper marketing in order to make their products more attractive to caterers.

**Box 5.1. Oregional.**

Oregional is a regional cooperation between various producers from a specific region in the Netherlands (Oregional is a project of the organisation Landwaard which is subsidised by the EU). It organises the supply of regional produce to care homes/clinics, schools, caterers and consumers. It also organises and promotes touristic/leisure activities in the countryside and surrounding farms. Oregional thus functions a link within and organiser of 'new product-market combinations', as it calls them.

Producers working within Oregional have to meet a number of criteria on aspects such as countryside conservation, environment, animal welfare, greenhouse gas emission, water management and education. Producers receive points on each of these aspects, and a minimum number of points are required in order to belong to Oregional. In addition, producers can earn extra points on aspects on which they score high (for example, one producer may do much with animal welfare, while another engages in nature conservation). In this way producers' strengths are recognised and encouraged.

Participation within Oregional offers producers a minimum of 15% more margin on their products compared to conventional channels. The cooperation organises the procurement of products, the necessary processing of produce and their sale and marketing. It also organises the financing of such supply chains. The primary difference between Oregional and conventional wholesale is that the focus is on setting up supply chains confined within a specific region, and that producers are given a more prominent role and more visibility. Also a greater emphasis is put on sustainable production and the link between producers and consumers.

According to Oregional many restaurants and catering businesses recognise the quality advantage of regional produce and use it to add value and distinction to their services. One of the attractive aspects is the price/quality ratio. For a number of products, Oregional can offer products for about the same price as offered by 'conventional' wholesale.

There are various initiatives in the Netherlands which work on improving regionally based supply chains. The on organised by Oregional also works on improving the availability of local/regional products to caterers (Box 5.1).

### *5.5.4 Gaining more control over the supply chain; AFN's and catering*

Caterers can take a more pro-active position when it comes to procurement and sourcing. In the Netherlands Hutten Catering is such as example. Hutten Catering is hired by various Dutch organisations in the private and public sector. This caterer wishes to gain more control over where ingredients/food products come from and how they are produced. This offers more knowledge about production processes and inputs, as well as the ability to be transparent to their customers and clients. This is achieved through more personal involvement and contact with suppliers and producers. Hutten does not only wish to find out more about how food is produced and processed, but is actively involved within the supply chain in order to secure certain standards of production and provisioning.

Hutten Catering's has a 'from farm-to-fork' catering strategy (Box. 5.2). Part of the fresh produce used by Hutten is sourced amongst a regional network of suppliers (primary producers, wholesalers, processors). It looks for sustainable production criteria to be met and works closely together with farmers, processors and other experts to develop production further along these criteria. For example, Hutten works together with farmers and vets on making improvements in the use of certain medicines in animal rearing. In this way it embeds its business within a network of people, knowledge and skill, of local and regional farmers, food producers and processors. It is developing an alternative food network (AFN) (Renting *et al.*, 2003). What makes this particular AFN unique is that it is linked to the business catering sector, and covers all aspects of the food chain spectrum; food production, processing, preparation and cooking (culinary quality), and food hospitality and serving (inside the locale of the canteen). Furthermore, this AFN is made visible and explicit to consumers and customers (contract-lenders) by, amongst others, their own label (De Guijt). Via a special website customers can read about the De Guijt label, what is stands for, and the farmers which it works with. In this way Hutten is a good example of an AFN which is active in terms of creating an alliance with various actors within a food chain, embedding the food chain within a region, but also in terms of certain values (relating to both the culinary and the sustainable) and employing a clear marketing strategy through the De Guijt label (Roep & Wiskerke, 2012).

What makes Hutten an interesting example is that this AFN is embedded within the catering business, thus relating to the work-place 'fork' and not the 'home-situated fork'. Hutten produces most of their own food in one central kitchen. Being a relatively large catering business this is fairly unique, since many contract-caterers only engage in cooking to a limited extent. Thus, Hutten is not just busy developing qualities related to production methods, but also culinary qualities. Also, this business strategy affords Hutten a certain position within the catering sector, distinguishing it from other caterers. Instead of competing on price for instance, Hutten Catering looks to offer something other large caterers might not be able to: fresher foods, cooked in their own central kitchens, with a level of transparency in terms of the origin of ingredients that other caterers cannot provide easily.

### 5.5.5 Good canteen management, good kitchen

Realising sustainable canteen food provisioning requires good canteen management, specific knowledge and skill is needed to adapt sourcing, food preparation and menu content in such a way as to make sustainable provisioning possible. Caterers mention such aspects as regional/seasonal sourcing (as discussed previously), working in an efficient and cost saving way in the kitchen, cutting down on food wastage, adapting and creating new recipes, adapting menus and menu cycles. Since sustainable food alternatives are often more expensive such strategies are needed in order to manage costs. Various caterers indicated how the higher costs of sustainable produce could be compensated by saving money elsewhere, through a different way of working and realising efficiency in other areas in the kitchen. 'Cooking with organic potatoes and vegetables is more expensive, but by buying in regionally, sticking to seasons and by being creative and innovative in the kitchen, we can make it affordable. For example, we are looking at how to use all of the broccoli, the stalk well as the florets' (interview). In addition, costs are all about cost-perception.

**Box 5.2. Pro-active in the supply chain: the case of Hutten Catering.**

Overall Hutten's philosophy connects sustainability to food/culinary quality, where a major focus is on reducing food miles and the support for sustainable food production in the region where Hutten is based. Hutten Catering is in the process of setting up an internal certification system which formulates different criteria according to four aspects: freshness, artisan production, sustainability and healthy food.

Hutten aims to arrange its services in such a way that the distance between producer and end-user is never more than 300 km, although this is not possible with all products (e.g. with fish, in this case Hutten keeps to MSC certified fish). The products which are offered according to this sustainable sourcing and provisioning policy is represented by Hutten's label 'De Guijt', and Hutten aims to have 80% of its services fall under this label. The label represents a number of criteria on a number of topics, like for instance energy use, use of pesticides and food miles. On food preparation Hutten has a health policy, whereby recipes are constructed in such a way as to reduce salt and harmful fats, conserving nutritional value and discard E-numbers and other artificial additives. The 'De Guijt' line includes soups, salads, meals, sauces and a range of deep fried snacks. By cutting out a proportion of the middlemen, these products are not much more expensive.

Transparency within and knowledge about the supply chains from which Hutten sources is considered one of the most important aspects of its sourcing strategy. Hutten aims to gain insight into each step in the food chain of different produce. For example, of his pork supplies he knows who the primary producers are, who slaughters the pigs and processes the meat (see figure below). Also, for each of these links in the chain Hutten has different criteria (relating to quality, sustainability, etc.). In the case of pork for instance two criteria are that pigs get good feed and not too much medicine, so that the meat is of better quality and contains as little medicine residues as possible. This process of applying sustainability and quality criteria to different links in the supply chain is something Hutten aims to carry out for other products as well. Hutten indicates that this process is easier for some product groups than others. With bakery produce for instance regional sourcing is much more difficult since regional/nationally sourced flour is more scarce and more expensive than flour that comes from Russia and Eastern Europe.

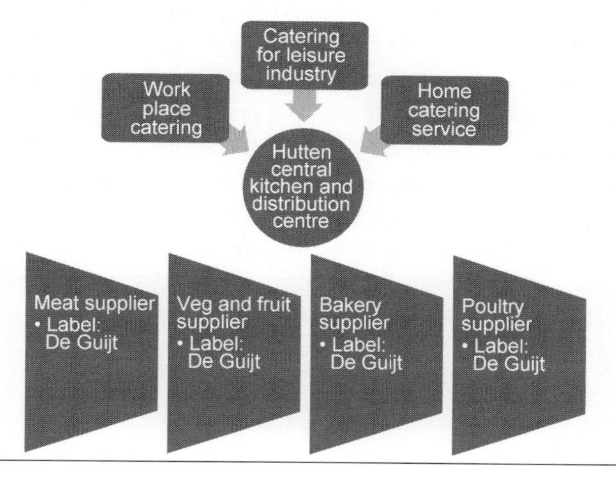

Skilled food professionals and managers recognise this and look to create customer satisfaction and favourable perceptions towards the catering service. An improvement of the quality of canteen food provisioning can offset limited increases in price.

Good canteen management requires skilled, creative and motivated staff. This applies to the managers, but also to the cooks and food preparing staff, and those serving the food. For instance, one project which looked to improve the catering facilities of an elderly home[56] found that the serving staff needed be brush up on their hospitality skills in order to create a better service for customers (Brok & Gorselink, 2010). A study found that the food intake of elderly persons within care homes could be improved by paying more attention to hospitality and food experience, and cooking more tasty better quality meals (Brok & Gorselink, 2010).

The extent to which caterers can be creative and find day-to-day solutions for sustainable provisioning is also influenced by the presence, or absence, of good kitchen infrastructure as we have previously discussed. Box 5.3 gives an example of how creative canteen management and kitchen infrastructure play a role in the transformation of a canteen in a municipality building of the city of Copenhagen (in-house catering) (Dogme2000, 2007). A number of structural changes were made as to the running of the kitchen, the way food was presented in the buffet area, the manner in which food produce was sourced and prepared, and the recipes used. It required a learning processes amongst canteen staff in terms of acquiring knowledge about organic and seasonal produce, preparing food with more vegetables and legumes, but also learning in terms of dealing with how to change food culture. Canteen serving staff also needed to learn about the changes made and the kinds of foods on offer.

### *5.5.6 Employing a 'food guru': the role of a good chef*

Hutten catering and the canteen in Copenhagen are good examples of cooking skills and knowledge playing an important role in the success of creating more sustainable canteen provisioning. The flair, knowledge and skill of food professionals is an important in part of good canteen management. The following case shows how the expertise of a head chef can be put to use and play a leading role in realising a sustainable canteen concept.

Albert Kooy is an established Dutch chef who has published cookbooks and is involved in the Dutch gastronomy world. His extensive knowledge of gastronomy and food was put to use in order to transform the catering services of an educational institution in the Netherlands. A number of years ago he transformed the catering services within a Dutch college, setting up a coherent and well-presented canteen concept (in-house catering system). His working methods are similar to those employed by Hutten Catering, as he works more closely with producers and suppliers according to a number of sustainability and healthy-eating criteria. In that sense he does not just 'buy' the products and ingredients needed for his daily service, but engages in a dialogue with his suppliers on food quality and sustainability issues. Kooy creates recipes with sustainability and health aspects in mind, looks for specialist suppliers, and works together with and producers to create recipes. In this way he creates a certain level of transparency about his supply chain and the ingredients used in his meals (Box 5.4).

[56] The project looked to introduce better quality, organic meals.

**Box 5.3. Sustainable catering in the city of Copenhagen, Denmark (Dogme 2000 programme).**

The Dogme 2000 programme is a cooperation between 7 Danish municipalities, working together to realise sustainable public procurement. The city of Copenhagen is one of the participating cities with a goal to realise 75% organic catering by 2008. The municipality counsellor in charge of environmental policy was a driving force behind the adoption of sustainable catering. He managed to get the city council to invest in sustainable provisioning.
Amongst the canteens that fall under the municipality are day-care centres, schools and sports facilities. In total there are 1,200 municipality kitchens, as they are termed. The plan of improving organic provisioning goes together with the goal of improving canteen food quality and nutrition. For Copenhagen, schools organic canteen provisioning is accompanied by an education package which aims to improve children's eating patterns, knowledge about food as well as learning certain skills.
The Dogme 2000 strategy is based on the idea that by changing consumption and provisioning practices the introduction of organic food can be realised without making the food services more expensive for end-users. The city of Copenhagen invested in the education of canteen cooks and staff, as well providing qualified counselling and consultancy in order facilitate the reorganisation of the canteens toward a better organic food supply. The changes made had to be cost-neutral, which meant canteen managers had to find creative ways to rearrange canteen provisioning. This was realised by:
- critically reviewing and renewing the menu plan;
- buying more seasonal fruits and vegetables;
- using more vegetables and more bread;
- baking their own bread;
- using more bulk foods, like pasta, rice, beans and lentils;
- using less meat;
- discontinuing unnecessary products;
- choosing recipes which use more vegetables;
- improving culinary standards;
- better using procurement contracts;
- reducing food wastage;
- carefully managing canteen economy.

Consultants came to the canteen to help realise these goals and find solutions in food preparation and procurement. This included explaining the principles of organic production to canteen staff members and finding out which products should be bought where and how everything could be realised within the existing budget. For example, costs were saved by baking bread themselves, which was cheaper than buying organic bread. In addition, the presentation of food was improved. For example, a wider range of salads became available in the canteens, which are served in large bowls on the buffet.
The city of Copenhagen does not have an official procurement policy stating various formalised procurement criteria. However, monitoring is in place, using a special computer model. Data from all the participating kitchens is collected and the model keeps track of how much organic food is served at each location.

**Box 5.4. Sustainable catering in 'Canteen'.**

The catering services at Van Stenden College in Leeuwarden are run by chef Albert Kooy. He has set up a concept for a sustainable canteen where provisioning is arranged according to the following aspects:

- The idea of having 80% vegetables and 20% meat instead of the other way around. By paying more attention to the vegetables, meals can become more sustainable. Pork and poultry are used more than beef.
- Serving a large proportion of organic food.
- The food served should be free of E-numbers and MSG. Such artificial additives are present in many products offered by catering wholesalers and suppliers. Kooy looks for suppliers who offer foods without such additives.
- Fruit and vegetables are kept seasonal as much as possible (e.g. in the winter white cabbage salad is served with a hamburger instead of tomato and lettuce).
- Development of own recipes and in some cases food products are made by an external producer according to these recipes (e.g. in the case of pesto and croquettes). For the foods that he makes in the canteen kitchen, he has selected producers who use methods which are more traditional/artisanal. The baker who supplies the bread for instance uses more traditional backing methods without the use of additives like flour improving agents, etc.
- Except for an organic coke and a kind of lemonade, all soft drinks have been discarded. 'Water-miles' and plastic are cut by serving tap water in glass bottles and adding gas for fizzy water, which comes with bottled water.
- Mono-packaging is not used and the canteen makes use of recycled/recyclable disposable cutlery and plastic.

Part of his canteen concept is to cut down on the meat component in his dishes and focus more on vegetables. Kooy wants to do something about the tendency within gastronomy to serve a lot of meat with vegetables as a side dish. His works towards making vegetables more interesting, and become a more central part of the menu. His sustainability policy centres around cutting down on food miles and serving more food that is in season. Being a chef he is interesting in creating a certain standard of food quality and taste. This is realised by using seasonal produce and producers and processors which work in a highly skilled manner and/or with more artisan production methods. He avoids the use of products which contain MSG as he believes it to be unhealthy and distorts the taste of food. He finds it important that young people especially get to know food that is made from scratch and with natural, fresh ingredients. Sometimes there are customers who have to get used to meals and foods which taste different then they are used to. He gives the example of some of the bread served in the canteen, which is made by a skilled, quality baker and different from the bread often sold in the supermarket.

## 5.6 Discussion

Caterers are sometimes criticised for their lack of creativity and innovation in realising sustainable provisioning. However, we have seen that caterers themselves indicate that their dependence on the terms of contracts, contract-lenders wishes, kitchen and canteen facilities, as well as attitudes towards catering can function as limiting factors toward realising more sustainable food supply. In order to realise innovation for sustainability, organisations (i.e. employers/contract-lenders) need to create an environment of incentive for caterers to be creative and innovative, invest in skill and knowledge of sustainable procurement and/or hire caterers who are skilful in sustainable food provisioning. Organisations need to see the advantages of good catering services within their organisation, and not just view catering as a cost-burden. On the other hand, caterers need to operate more as 'innovators' and less as 'followers'. Without actively investing and setting up their own sustainable catering strategy they remain dependant on the initiatives of contract-lenders and on the status quo, i.e. mainstream demand and conceptions about sustainable food alternatives. For most canteens it seems that working in a highly standardised way, keeping canteen food provisioning as constant as possible, together with a lack of professional skill, are factors which act as barriers to more sustainable food provisioning.

Innovation also depends on the dynamic between caterer, contract-lender/employer and the end-user. Experts indicate that caterers need to focus more on fitting provisioning to end-users wishes and needs, a topic which will be further explored in Chapter 6. Canteen food culture was mentioned as an important factor by a number of respondents. This 'canteen food culture' includes such aspects as end-user's expectations, the inflexibility of end-users food routines and the characteristic Dutch lunch preferences, but also catering providers' own expectations of end-users. Various caterers mentioned how changes made in provisioning can meet resistance from end-users and how caterers 'fear' this 'inflexibility' and the loss of demand as a result. Not only do end-users' routines and expectations play a role, but also caterers' own interpretation/prejudices of these issues and their how their customers will react to change. Chapter 6 will consider this topic further, starting from the idea that food culture is something which comes about in the interaction between provisioning, providers and end-users, within the context of the canteen locale.

Generally, it seems that caterers see sustainable canteen provisioning in the light of adding value to their services. In other words, food sustainability is inherently connected to and cannot be seen separately from, aspects of food and service quality. Food professionals achieve sustainable food provisioning by offering 'better' catering services in aspects of better quality food, in terms of taste, enjoyment, authenticity, etc. From the perspective of a food professional who is offering a service this is a natural vista through which sustainable provisioning is viewed and where the opportunities for realising sustainable provisioning are sought. Some caterers, for instance contract-caterer Hutten Catering, have developed a catering service where sustainability has become an intricate part of the business structure. Here values of sustainable production (environmental aspects, animal health and welfare aspects) are interwoven with values on culinary quality, integrity and authenticity to take up a certain position within a market, to be able to build a business which offers something other caterers do not or cannot.

### 5.6.1 Creating contracts which serve sustainability

Although the issue of contracts and contract-setting has only been partially studied in this research it has shown the importance of contracts in determining what kind of food will be served in canteens. Resistance to sustainable development in canteen food provisioning can be caused by the inflexibility which arises from being tied to a prior contracts. Especially the larger contract-caterers may experience inflexibility due to prior arrangements and long-standing business relationships with wholesalers and suppliers and find themselves unable to quickly switch suppliers. Larger contract catering companies often operate within a more rigid system of standardised food provisioning and contractual relationships compared to smaller independent caterers. This can make it more difficult for catering managers on the ground to change to other suppliers and make changes in menus, thus making them less flexible to adapt to direct cues and patterns of consumption of customers. Structural changes in food policy need to be decided by the head office first. In such a system it is vital that sustainability criteria are built into the individual contractual agreements with contract-lenders from the start. The caterer and the contract-lender must make clear agreements on how sustainability is to be realised.

Smaller caterers may have relatively more freedom to change suppliers, which also allows for the opportunity to try out various different suppliers (see also GM&FNE, 2007), and are in a more flexible position to make 'on-site' changes, i.e. changes in recipes, make special offers and/or introduce new products. However, large contract catering companies have the advantage of a more powerful position within the market. The large players in the catering sector are multinational corporations which have the potential power to demand certain products from large food suppliers, and even to effect change in more international supply chains.

When we talk about canteen food consumption we need to keep in mind that 'consumer choices' are made at two instances; at the time of setting up contracts with caterers/food suppliers and the moment end-users make use of catering services. Especially in hired catering (contract-catering) taking up sustainability criteria within contracts with caterers is crucial, especially where contracts are long (5-6 years). Catering contracts must be set up in such a way as to promote and enable sustainable provisioning. If sustainability is to be realised it needs to be taken on board from the early stages of the whole procurement and contract process and built in solidly within the details of the terms of reference. At the same time contracts should not stifle the development, innovation and creativity of the catering service.

A range of different criteria can be taken up in a contract, i.e. from environmental to social, and socio-economic. The Dutch procurement criteria for sustainable catering (set up under Dutch government initiative) and the EU guidelines give ample indication of the kind of sustainable criteria which might be employed. Overall, the issue is not so much which criteria can or should be used, but how to include these criteria, that is, how to formulate sustainability criteria and goals in such to realise the kind of sustainable provisioning preferred. Criteria with reference to the use of local/regional supply chains remain contested since they may conflict with EU goals of non-discrimination. However, when near-sourcing is phrased in terms of various quality aspects (e.g. taste, freshness) it conflicts less with the principle of non-discrimination. This 'creative procurement' can go a long way to encapsulate different requirements, translating sustainable procurement to the issues which are considered important by any given organisation, i.e. contextualising procurement.

Contextualisation also offers the opportunity to connect procurement to shared food values and/or other sustainable development goals within an organisation. Using procurement criteria and contract setting in such a way as to realise ones sustainability goals/wishes requires skilled catering and facility management. Especially when dealing with larger contract and the complex matter of EU tendering. Various studies show the importance of skilled professionals playing a key role in implementing sustainable procurement criteria (Adrichem, 2009; Eurocities, 2005; He & Mikkelsen, 2009; Morgan & Morley, 2002).

### 5.6.2 The presence of a real kitchen

We have seen that kitchen infrastructure has an important role to play in determining what kinds of options are available to those who want to 'green the canteen'. Canteens with skilled food professionals and proper 'cooking' kitchens will be better equipped to make changes in recipes and menus themselves. This allows for a degree of creativity and flexibility in realising sustainable provisioning. In cases where kitchens are absent caterers are more dependent on the availability and quality of sustainable food alternatives from suppliers and wholesalers. In other words, here canteens are more dependent on the market-availability of sustainable pre-prepared meals and meal-components (e.g. soups, salads, full warm-meals) within the food chain. Considering that freshly prepared, good tasting meals are areas where caterers can add value to their services (see also Chapter 6), not having a kitchen seems seriously limiting.

### 5.6.3 Near-sourcing

During the research it appeared that near-sourcing and seasonal scouring often has a practical reason. Near-sourcing in many cases is not a goal in itself, but rather plays a role in enabling sustainable provisioning cost-wise, and in adding value to catering services. Many food professionals (cooks, caterers) engage in near-sourcing due to certain quality advantages like freshness and taste, or making the use of organic or other sustainably produced foods affordable. Others want to support their local economy, and see socio-economic benefits in local/regional sourcing. Some speak about the importance to know the origin of their produce, to be able to offer their customers a degree of authenticity. It allows them to gain more control and transparency on food supply (food origin, production methods).

These aspects should not be forgotten in the debate on whether or not local/regional sourcing should be included as a sustainability catering criteria. The scientific and policy debate on local sourcing is often rationalised and argued in terms of the interests of North and South, or in terms of $CO_2$ emissions. The rationale of food service providers lies within a practical reality; near sourcing can offer food professionals practical solutions for improving their services and realising more sustainable provisioning. Many will still use banana's, coffee and pineapples. But sourcing their vegetables (for example) locally can make using organic produce affordable and help support neighbouring producers. For food service entrepreneurs supporting the local, sustainable economy might also be acquiring a certain meaning in light of the current context of the economic crisis and the loss of business for farmers and other food producers in the Netherlands.

### 5.6.4 Competence

Realising sustainable food procurement requires good canteen management. Competence, having the right knowledge and skill is a reoccurring theme within this chapter. It seems that realising sustainable provisioning has as much to do with the availability of sustainable food alternatives on the market as with the competence of those in charge of organisation catering services. As we have seen above, knowledge and skill is required for the setting of procurement policy and sustainability criteria, but it is also needed 'in the kitchen'; on the sourcing of foods, constructing menus, food preparation and composition. And, it is needed in order to realise sustainable provisioning without unrealistic increases in costs.

Here we discover the importance of the role of food professionals and their capabilities. Whereas organisations (the caterers' employer in the case of in-house catering and the contract-lending party in the case of hired catering) need competence to manage catering procurement processes, capable food professionals play a key role in providing the actual service to the end-user. This may require a process of learning and/or hiring/involving experienced procurement officers, managers or food professionals (cooks) specialized in sustainable catering (see also studies by Adrichem, 2009 and Heerkens, 2009).

### 5.6.5 Further study

The potential role food professionals could play in the sustainable development of the agro-food sector and everyday food consumption practice(s) has been largely ignored by researchers and policymakers. Further research might be able to bring more clarification as to how organisations can improve sustainable procurement and contract setting processes. Food professionals are taking a more active role in initiatives to realise both more sustainable and better quality food service. Research could consider how to strengthen the competences of a range of food professionals, e.g. catering managers, cooks, facility managers to help realise such goals. Some respondents have suggested that in the Netherlands many facility managers, cooks and caterers do not have adequate knowledge about sustainable procurement and food (e.g. food seasonality, culinary skill and knowledge). Others have mentioned the lack of skilled cooks working within work-place catering generally. Thus one area of investigation could be to find out if this is in fact true, and consider how the education programmes for facility/hospitality management and cooks might be improved. The food provisioning and infrastructure of the 'public plate' is another area deserving attention. How can healthier and more sustainable food consumption practices be realised here? The 'public plate' relates to food provisioning within primary and secondary education, hospitals and care facilities, and areas within the (semi)private sector, e.g. sports centres.

Furthermore, it would be interesting to dedicate more research to place-based AFN's linked to the restaurant and catering sector; alternative food services networks. For example how can the public plate benefit from connecting to local and regional food producers, and vice versa. What role can strengthening the local/regional food infrastructure/chain mean for realising interrelated policy goals, like sustainable agriculture, healthy eating, etc.

# Chapter 6.
# Sustainable catering and the end-user

## 6.1 Introduction

In the previous chapter we discussed initiatives and experiences in realising more sustainable food provisioning in work-place canteens. In this chapter we focus on what is happening within the locale, or consumption junction, and analyse more closely the social practice of eating in a work-place canteen from the perspective of the end-user. Sustainable canteen provisioning often goes hand in hand with processes of adding value to catering services, where the focus is on food quality and food experience. In the previous chapter we saw that changing food provisioning within a canteen can mean a confrontation with the existing 'canteen food culture', that is, with end-users habits and routines, their expectations and standards concerning canteen food services. The main objective of this chapter is to understand what kind of approach could be taken towards the end-user and what strategies might be employed to encourage sustainable food consumption practices within an organisation.

In terms of our conceptual model (Figure 5.1) we are approaching the practice of 'having lunch in a canteen' from left to right. Our first aim is to gain a better understanding of the canteen practice from the perspective of the end-user; the routines and considerations of consumers within this work and study related setting and the position of the end-user (employees and students) in relation the organisation of canteen food provisioning. The second aim is to understand what is happening within the canteen is a locale, a place where food culture(s) are manifested. What kinds of strategies can be employed by caterers to encourage sustainable food consumption practices.

Following Section 6.2 on methodology, we first discuss Dutch canteen food culture, and some of the main trends in canteen food practices (Section 6.3). We then consider the social practice of having lunch in a canteen in some more detail, discussing amongst others to what extent sustainability does or does not play a role in this setting yet (Section 6.4). This is based on the focus group with end-users; employees and students from a Dutch university. Then we discuss how end-users might be targeted and/or involved within sustainable food procurement and provisioning processes and to what end, based on interviews with catering experts and literature (Section 6.5). We explore different strategies that can be used to encourage sustainable food consumption on the canteen floor (Section 6.6) based on the focus groups with end-users and the one with caterers, as well as the visits to various innovative canteens and caterers. The last section then presents a discussion on the findings (Section 6.7).

The research questions addressed in this particular chapter are:

- What kind of approach toward and level of involvement of end-users would enhance sustainable catering?
- Which (future) food provision strategies are particularly suited for encouraging sustainable food consumption in the canteen?

## 6.2 Methodology

This chapter is based on the same set of empirical material as used for Chapter 5 (Section 5.2 and Figure 5.1). However, in this chapter we discuss the findings which focus on the end-user and the locale, as formulated in the research questions for this chapter.

Two focus groups were held, one with caterers and one with end-users. Section 6.4 covers the results from the end-user focus group, where participants were asked to discuss their everyday lunch practice, their food choices and thoughts on sustainable canteen food provisioning in their canteen. Section 6.6 discusses findings from both the end-user and the caterer focus group on how to encourage sustainable food choice within the canteen. In both focus groups a short presentation was given on various strategies to encourage more sustainable food consumption in the canteen setting. After the presentation, the question was posed: 'How do you think consumers should be encouraged to make more sustainable food choices in the canteen?' Participants were divided into groups and each group was given one strategy to discuss.

Participants for the end-user focus group were recruited from a Dutch university. This seemed interesting as there would be a mix between younger and older participants (employees and students). This particular university was chosen because the in-house caterer is relatively active in sustainable canteen provisioning, providing both organic and Fair Trade food products and engaging in near-sourcing. Although no prior selection of participants was made, the people that participated appeared to be relatively environmentally conscious, either because of their working background (working in environmental sciences), or because they indicated this during the focus group.

## 6.3 Canteen food culture: the main trends

Generally speaking, Dutch canteen food culture is associated with having a quick, affordable bite to eat. In the Netherlands lunch is 'fast' and light, with foods like sandwiches, soups, fruit and dairy (glass of milk or bowl of yogurt) being the most characteristic ingredients (Foodstep, 2005). Canteens within the work-place, schools and organisations for higher education do not have the tradition of providing full warm meals, as it is the case for instance in Scandinavia. This quick and light lunch culture is reflected in the kinds of food offered in work-place canteens. Prominent in canteens are pre-prepared food products and 'ready-to-eat' food products, like fruit and dairy snacks. Another feature of Dutch lunch culture is that on average the Dutch are used to spending relatively little money on their lunch (2-3 euro's).[57] This may partly be explained by the fact that around 46% of Dutch employees still bring sandwiches from home and then buy something 'extra' in the canteen, such as soup and something to drink (Foodstep, 2005). However, according to catering sector consultants Van der Meer (2010) this could be expected to change in the near future. Young people, and thus future generations, are said to be more familiar with using out-of-home food servicers (meals and snacks), and more used to their concepts and prices. For this reason, young people are seen as an important target group for the catering industry (Van der Meer, 2010).

[57] Http://tinyurl.com/l2pfgud.

Food quality and enjoyment, as well as healthy food (more fresh vegetables and less harmful fats, sugars and salt) and sustainable eating (especially organic food) appear to be trends within the catering sector. In the past, studies have indicated that canteen users would prefer a more restaurant-like feel to canteens, and certainly in the past years caterers have paid more attention to serving more tasty, freshly prepared foods, with more salads and vegetables (CGE&Y, 2002; Foodstep, 2005). Successful caterers are said to be those who offer good price-quality ratio and work with more restaurant-like concepts (Van der Meer, 2010). Out-of-home food expenditure is expected to rise over the coming years, and this may be a positive development for caterers operating within the settings of the work-place and educational institutions (HTC, 2007). Caterers might be able to benefit from providing the kind of food services which fit busy (work) lifestyles. For example, by providing more warm meals during the day, and in the evening, or take-home meals. Indeed, trend watchers have also reported that catering services are moving from business-to-business services in the direction of business-to-consumer services (idem). This means more actively designing catering services to fit the on-site target group.

## 6.4 The practice of having lunch in a canteen; focus group discussions

In order to gain insight into canteen food consumption practices, a focus group was held with end-users. The participants were students and employees of a Dutch university whose (in-house) caterer provides organic meals and food products, as well as Fair Trade labelled foods. The three main issues discussed with the participants were: end-user's lunch routines, their food preferences, their use of sustainable food alternatives provided in the canteen, and their opinions on sustainable provisioning in the canteen. From the analysis of the focus group transcript, a number of topics were selected (Table 6.1).

### *6.4.1 Lunch practices: routine and flexibility*

People's lunch practices appeared varied, some described it as a fairly fixed routine, whilst others described how they 'adapted' their lunch habits to their daily schedules. Most participants only spend half an hour on their lunch, and lunch practices seem to contain a mix of both routine and ad hoc elements. For some, having a canteen lunch has a social function, an opportunity to socialise with colleagues, usually at a fixed time during the day. However, others mentioned that

*Table 6.1. Main topics from the discussions in the end user focus group.*

| |
|---|
| Lunch practice: routine eating (same time, same place, same people) and flexible eating (fitting lunch to work schedule) |
| Value for money: compare & contrast service/products, comparing on the basis of experiences at other places |
| Sustainability issues: diverging opinions/ideas/definitions, often mentioned issue = waste and packaging |
| Sustainability and end-user role: 'canteen end-user is not an activist' |

they often have their lunch behind their computer. It seems that having lunch at work is about trying to be flexible, to 'fit it in' with work related patterns. Some explained that they do not take as much time for their lunch as they maybe should. How much time is spent on lunch, what foods are eaten and whether or not one eats in the canteen depends on the schedule of the day, for example on having appointments, classes, etc. Indeed other studies have shown this same element, that time-constraint is a factor in having lunch within the workplace (Spaargaren *et al.*, 2013) and that it can even be a reason not to visit the canteen at all (Foodstep, 2005).

The lunch-content of most of the focus group participants fits the stereotypical model of the Dutch lunch which consists of bread (sandwiches, roles/baguettes), soup, milk (or other dairy drinks, e.g. yoghurt drink, buttermilk) and sometimes a deep fried snack. A number of participants mentioned that they eat home-made sandwiches for lunch. Soup and dairy drinks (milk/yoghurt/buttermilk) are bought in the canteen to complete the lunch. A couple of participants, however, said they buy a warm meal from time to time, or sometimes treated themselves to pancakes or fries. There is certainly room for whim and 'simply choosing what 'looks good' and what one 'fancies' eating' (comment participant focus group). One participant commented that his food choice was sometimes dependant on 'trivial' factors such as the size of a queue.

### 6.4.2 *Value for money*

Another major topic within the focus groups discussion related to the issue of getting 'value for money'. Firstly, various participants seemed to expect that canteen food is inexpensive. This may possibly have to do with the fact that in the past Dutch organisations subsidised their catering services. Participants associated canteen food with a social-service, i.e. not compatible with commercial food provisioning. This is reflected in a number of comments made by participants on the responsibility of the university to provide both cheap and sustainable food.

Secondly, value for money was argued in terms of quality aspects, which in turn was often reasoned in terms of comparison with prior canteen food experiences. For instance, students compared the canteen prices with prices at supermarkets, arguing that buying bread and peanut butter or cheese in the supermarket around the corner was much cheaper than buying sandwiches in the canteen. 'If you buy a sandwich here [referring to the canteen] then you pay 1.50 euro, but when you buy a loaf of bread and a pot of peanut butter [from the supermarket around the corner] you can eat for a week' (quote from focus group participant). Other participants made various comparisons between foods. For instance, buying a warm meal at the canteen was considered much better value for money than buying sandwiches, which can easily and cheaply be brought from home. 'It depends on what you compare it to. In the case of hot [canteen] meals it is cheaper...' (quote from focus group participant). Another participant mentioned that when food like pancakes were on the menu he was more likely to make a purchase in the canteen.

Comparisons were also made using prior experiences with similar canteens as a reference. Two participants compared the university's canteen provisioning to canteens at other universities they had visited or worked at. One participant mentioned that the offer in sustainable foods was much better at her previous job, and therefore her view of this canteen was not as positive as it could be. 'I'm always comparing it to Germany, at [the university where I got my degree] I paid

50% less then here in the Netherlands and there was even much more organic food' (quote from focus group participant).

### 6.4.3 Sustainability and the role of end-users

The discussion on sustainable food provisioning revealed a range of different opinions. Some participants demanded all the food to be organic, as then it would be more easy to choose 'the right' product. Others thought organic food in would make canteen food too expensive, whilst others pointed out that it was good to spend some more money for organic food.

A topic which came up during the focus group and dominated the discussion on canteen sustainability was waste from food packaging. Plastic packaging appeared to be an important factor for the end-user participants in judging how sustainable the canteen is. 'In the old days the slices of cheese were not all packed separately. These days it's all packed separately, the jam, the chocolate sprinkles and the hot sauce...' (quote focus group participant). Another participant responded with 'That annoys me too. I am disappointed, packaging is a real issue, in the canteen, but also in all the catering. For instance, catering during workshops. They package everything.' And still another group member argues: 'I had lunch at another university the other day and there it was even worse'. This is in contrast to the reasoning within the food (catering) industry that packaging prevents food wastage and therefore has an environmental benefit.

Most of the participants knew what kinds of sustainable foods were available in their canteen (Fair Trade products and organic food). Each participant mentioned choosing sustainable products in the canteen at one time or another, but not a single participant expressed real structural or routine use of the organic, Fair Trade or vegetarian foods offered in the canteen setting. This in spite of the fact that generally the participants found sustainable food consumption a very important issue and were in mutual agreement that sustainable alternatives should be available in the canteen.

One participant indicated that canteen food consumption is different from food consumption at home. At home sustainable alternatives like organic food were bought often, but in the canteen this was different. He pointed out that in the canteen one is dependent on what foods and services are being offered at any particular moment. In fact he felt he had little influence on what is provided by catering services. 'Most [canteen] consumers are not activists. At home I only buy organic products, but here [at the university] I have to buy what is offered. There are not very many people who comment on how things should be done differently ... As a user you always feel as though things are not possible when you ask for a change, that's why you just don't bother to ask any more'. Another participant had a different opinion, he said: 'I think especially [the canteen] is an example of a place where employees have a choice. Paper and computers are ordered for at the central level of the organisation, we don't really have a say about that ... [The canteen] is the only place where people do have a choice ...'. Overall, however, there seemed to be agreement in the group that it is primarily the caterer who is responsible for sustainable provisioning of the canteen. The university administration or the end-users themselves were not specifically mentioned in this respect.

## 6.5 Involving the end-user in canteen provisioning

### *6.5.1 The captive consumer*

The canteen customer is sometimes described as a 'captive consumer' and indeed as we have seen some focus group participants expressed they felt a certain degree of dependence on and restricted influence over what is served in the canteen. The way in which many canteens are organised probably enforces this kind of passivity. From the literature, the interviews and the focus groups it appeared that canteen end-users are hardly involved the decision making processes and organisation of canteen food provisioning within the organisations where they work and study. In some organisations a consumer panel or consumer committee is in place to represent the interests of end-users. These are sometimes consulted by facility managers and/or caterers in the procedure prior to closing new contracts with caterers. However, one study on catering in higher education found that caterers hardly communicate on a regular basis with end-users, for instance via an end-user panel (Van der Meer, 2010). It also revealed that on average the contract-lender is more satisfied with the hired catering services than the end-users (Damen & Van der Meer, 2010).

The same research also showed that although customer (end-user) satisfaction surveys are popular, end-user satisfaction/perception is often not used as input for developing new strategies and policies, or used in the process of procurement, tendering and contract setting. Also, end-user satisfaction does not often play a formal role in determining whether contracts should be extended (Damen & Van der Meer, 2010). In addition, 'service level aspects' such as expenditure patterns of end-users sometimes hardly influence contract agreements, something which is detrimental to service quality (Damen, 2010a). Overall catering within higher education is characterised by a relatively stronger business-to-business relationship compared to a business-to-end user relationship. The lines of contact and communication between the canteen providers are better than between providers and end-users.

More attention to and/or more involvement of the end-user has been named as an important factor in making catering services a success (Damen & Van der Meer, 2010; Van der Meer, 2010). It is needed to realise a better fit between canteen food provisioning and the target-group, something which can be equally relevant for processes of organising more sustainable food provisioning. In Chapter 5 we mentioned how caterers in interviews and the focus group pointed to the resistance from end-users, when trying to introduce more sustainable food alternatives in the canteen. 'Canteen food culture' can provide a barrier to changes in food provisioning, in terms of the habits and expectations of both end-users and caterers (Section 5.4.4). These aspects might be dealt with through more appropriate and pro-active approaches toward the end-user. The following section explores this. It is based on literature research, interviews with catering experts (caterers, and a facility management consultant) and the focus group with caterers.

### *6.5.2 Creating a better fit between end-user and caterer*

In the catering sector, it seems that end-users are often being reduced to a category of 'demand' in terms of product sales. On the basis of the latter caterers make choices on which products to include in their assortment and what supply orders (and contracts) to arrange accordingly. If products in the canteen do not sell, a quickly drawn conclusion is that there is no demand for them. However, present demand for sustainable food in the canteen may not always be a good predictor of the potential success of sustainable provisioning. In addition, the danger of focussing too much on the paying customer, is that 'potential' customers, and 'potential' wishes of customers or latent demand are not sought out (Damen & Van der Meer, 2010). A more adequate feedback system can improve on the interaction between end-users and canteen food service providers (caterers and the employee/contract-lending party) and provide input for strategies of sustainable food provisioning and encouraging demand for such provisioning within the canteen.

One approach is to arrange for a more structural dialogue with end-users through setting up a regular communication structure in order to discover end-users' wishes, complaints as well as food sustainability concerns (issues important to end-users). One example of such a structure is applied by Hutten Catering. Hutten Catering uses experienced food professionals to actively monitor the developments at each canteen location. These so-called 'selected chefs' find out about the (changing) wishes of their clients (contract-lenders) and end-users. They then come together on a regular basis to discuss these aspects with the head of Hutten Catering. The idea is to come up with new concepts, products and/or services which are better adapted to the needs of various locations and their actors.

During the focus group with caterers participants were asked how they gather information on sustainability concerns and the wishes of their customers. Caterers indicated that communication with the end-user was mostly approached by talking to customers face to face, by customer satisfaction questionnaires, or using card or on-line systems where end-users can make complaints. One caterer explained that customer surveys only have limited value. He often found there to be a discrepancy between what people think/say and what people actually buy in the canteen: 'People can say all kinds of things, it does mean they will act on it]' (quote caterer focus group). Another caterer expressed the need to be very careful when asking about individual opinions since this might create certain expectations. Several caterers expressed that they preferred one-to-one contact between themselves and the end-user. Through personal communication questions could be dealt with in a more direct manner, and confusion or false expectations of customers could be prevented. Not only does one-to-one personal contact, end-user panels and surveying allow caterers to find out what end-users want and think, these instruments may also help to legitimise certain procurement choices under the employees of an organisation and create a firmer basis of support for procurement decisions within an organisation. Support amongst employees (who may or may not use the canteen) and other stakeholders can be very important for sustainable food provisioning initiatives to succeed (Adrichem, 2009; He & Mikkelsen, 2009) (also interview caterer).[58]

This support may be also realised by translating an organisations' food procurement policy (and/or the choice for a certain caterer for instance) to certain shared values or goals which reflect an organisation as a whole. A good example is that of a Dutch provincial government that was able to link its food procurement policy to the problem of regional ground water pollution

[58] In the project on organic public food provisioning in three Danish municipalities for example, He and Mikkelsen (2009) found that a lack of feeling responsibility for good school meals amongst pupils, teachers, other school staff and parents made changes in food provisioning more difficult.

caused by agriculture. This was used as an argument for choosing to serve more organic foods in their canteen. They communicated to employees that the organisation wanted to actively support sustainable farming methods. Thus, procurement policy was explained in terms of a concrete local problem with which employees could relate to. One focus group participant reported a similar strategy. He indicated that food procurement policy reflected company philosophy and values. This was communicated to employees thus linking these values to the food on offer in the canteen. 'One can communicate [why] the company finds these products important, that there are certain choices behind this type of procurement ...' (quote focus group participant).

## 6.6 Strategies for the canteen floor: end-users versus caterers

Above, we have touched on how to approach the end-user with respect to canteen food procurement and provisioning within an organisation. Targeting the end-user in context, i.e. on the canteen floor during lunch-time, is another issue. Here we encounter end-users who are engaging in their daily 'lunch moment', in the context of a (busy) work or study schedule, during 'the employers time'. Furthermore, we may be dealing with some general 'Dutch' lunch habits, relating to a sober lunch of relatively short duration, which should not be too elaborate and too expensive. With this in mind, we now turn to discuss what kinds of strategies might be employed in such a setting in order to make sustainable canteen food provisioning (or the introduction of more sustainable food provisioning) a success. First, we consider the discussions from the focus groups on this matter. Both caterer and end-users were asked to discuss various strategies for encouraging sustainable food consumption practices in their canteen. Secondly, we take a look at some innovative examples of such strategies based on visits to canteens and interview with caterers.

During both the focus group with students and employees from a Dutch university and the focus group with Dutch caterers, participants were asked to discuss four different types of strategies which can be used to encourage sustainable food choices within the canteen. These were presented as 'communication strategies' to participants in a power point presentation. These were:

- The labelling strategy: clearly labelling of food products like organic, Fair Trade, MSC fish, etc. or, creating a separate labelling system for the canteen for such foods.
- The information strategy: providing information about sustainable food alternatives provided in the canteen, on for instance packaging and/or posters.
- The 'story telling strategy': framing products through 'story telling', that is, providing factual information about the producer and production methods for example, in an attractive visual way (for example with pictures).
- The 'food experience' strategy: creating a better atmosphere in the canteen, improving canteen (service) attractiveness, serving different foods and meals (changing recipes), more culinary quality and food enjoyment.

The first two strategies convey factual information about the food and its origin. Offering labels and information implies sustainable consumption is encouraged via a more 'rational' route. The last two strategies are more to do with marketing and creating added value. The 'story telling' strategy is meant to 'personalise' and 'frame' the food. The food experience strategy is where attention is paid to the entire experience of visiting the canteen; the canteen atmosphere and the taste of the food. In other words, sustainable food provisioning is accompanied by changes in such aspects as canteen design, recipes, cooking methods and interaction between customers and catering staff. This was also described as a 'canteen make-over'.

### 6.6.1 End-user focus group

According to most participants of the end-user focus group labelling and information was considered the most promising strategy. Labels, as well as additional product information should be made visible and/or available in the canteen setting, so they argued. The three reasons for preferring these strategies were: labels would increase consumer awareness, it would allow for more freedom of choice and it would promote a discussion about sustainable provisioning within the university. After some discussion within the group during the session, it was agreed that there should not be too much information and labelling, the canteen should still be an inviting place to come to eat. 'With [information] signs everywhere this could give a chaotic impression and it could be interpreted as advertisement overkill' (quote end-user participant). Participants mentioned that the trustworthiness of the labels and the information was important. There were varying opinions on this issue, some thought it necessary to limit the presence of labels to only a few trusted ones, whilst others thought that consumers would simply get used to all kinds of labels over time. The 'story behind the food' approach was seen as a promising additional strategy to the more factual consumer information and labelling, but a story without an official label was considered untrustworthy. The food experience strategy was considered good for the social relationship between colleagues and worker wellbeing. However, it was considered too weak on raising awareness about sustainability within the canteen and for some it was associated with more expensive and luxurious food.

Participants had a short discussion about the extent to which end-users should be involved or informed about the sustainability of the canteen food. There did not seem to be consensus on this topic, opinions were very varied. Whereas one participant wanted all canteen food to be as sustainable as possible but not to be informed on the details, another person stated that this information would encourage him to visit the canteen and eat there.

### 6.6.2 Caterer focus group

Caterers more or less unanimously agreed that the most effective strategy in encouraging sustainable choices was the food experience strategy. The least preferable was labelling and consumer information. Food experience was a way to add and draw attention to aspects such a product freshness and taste. In addition, caterers were in agreement that the best way to promote organic products is not to frame it as environmentally or socially beneficial, but to make organic products appear 'less alternative', as one focus group participant put it. Letting customers taste organic food, for instance, was considered an excellent way to let them get to know new products and recipes. Labels were more relevant as to purely informing end-users as to the content of products and meals, but not as a strategy to promote sustainable food provisioning in the canteen.

One participant suggested that information on environmental benefits are uncertain and that claims made on the products in the canteen based on extra product information might be misinterpreted by end-users. Personal explanation about products and what they involve was considered more promising. This participant indicated that it is important that organisations explain how sustainable procurement fits into the company philosophy.

The focus group findings show a distinct discrepancy in preference and reasoning between caterers and end-users. The caterers preferred the food experience strategy whist the employees and students revealed more varying opinions on this topic but seemed to agree that labelling and information were the most promising. End-users expressed the need for trustworthy information provided by a well-know and trustworthy label for example (like the organic label and the Fair Trade label). However, various caterers participating in the focus groups and interviews indicated that when it comes to sustainable provisioning they want to refrain from being 'educators'. Canteen food service is about food enjoyment and good service, with a relationship based on trust. One respondent formulated it thus: '[it is]about having an undisturbed and enjoyable lunch', not 'bothering consumer with too much information' (interview chef caterer). Caterers fear that labelling and information might interfere with providing an enjoyable canteen experience. Indeed, findings from a Dutch study on the effect of $CO_2$ labelling on canteen food choice found that end-users felt they were being bothered by the presence of certain types of information tools on $CO_2$ output and made to feel guilty during their lunch (Spaargaren *et al.*, 2013). The study reveals that the design of the information, where and how the information is offered to the consumer, plays an important role in whether it is perceived as disturbing, and thus whether it is effective in encouraging a more sustainable food choice.[59]

## 6.7 Exploring the food experience strategy's potential to change food consumption practices

In approaching transitions in sustainable food consumption this study employs a practice approach, with the premise that changes in consumption patterns require (or result from) changes in social practice(s). The question is then which measures would be able to induce changes in (consumption) practices within the canteen? The 'food culture' within an organisation/canteen, people's habits, routines, tastes, preferences and expectations, may create a degree of resistance towards changes on food provisioning. We have also seen that although lunch time in a work-related setting is prone both to routine and to adapting lunch to work-schedules, it is also influenced by momentary considerations and expectations and standards related to prior visits to canteens. The momentary considerations are really about what is happing in the moment within the locale; what foods and meals are on offer, do they look good/smell good.

[59] Some information tools were considered 'annoying' whilst others were considered interesting or useful. It seemed that especially information given after lunch was more appreciated than information given during lunch (in the form of a film played in the buffet area during lunch-time). $CO_2$ labels showing 'grams of $CO_2$' did not provide end-users with a useful reference point. Something which both this study and the $CO_2$ labelling study show is that information is perceived and judged differently by different people. In the case of the $CO_2$ labelling project some focus group participants said they preferred simple labels with little added information, whilst others preferred more additional background information. Overall the study concluded that consumer information needs to be carefully designed to 'fit' the practice context ('having an enjoyable lunch at work').

Participants within the focus group spoke about introducing organic food and how this must go hand in hand with tasting such products, as well as providing both catering staff with information on organic production and possible even the producers themselves. Not surprisingly perhaps, most caterers who participated in some way in this research (interviews, focus group) mentioned taste and food service as a central to their strategy to take up more sustainably produced foods in their catering services. The same is true when we look at innovative caterers, such as Hutten Catering, the canteen at the municipality of Copenhagen (Chapter 5) and Albert Kooy's Canteen. For these caterers creating more sustainable canteen provisioning goes hand in hand with creating 'food experience'. These caterers are adding value through serving better tasting meals, more interesting foods and recipes, better service in the canteen, a better atmosphere in the canteen through improvements in the design of the space. The food experience strategy is nothing new, it is the core principle of being an entrepreneur in the catering and restaurant sector, but it is not often mentioned or studied by scientists and policymakers concerned with encouraging sustainable food consumption.

Albert Kooy's approach to food experience meant creating a canteen concept that would appeal to young people. His catering services are located within an educational institution which offers hotel management studies. His restaurant called 'Canteen' serves food for students and staff members and has a very modern feel whilst at the same time applying some traditional and sustainable principles concerning food. Tradition is seen in the use of authentic or artisan way of baking bread for instance. There is no use of artificial additives in the food and most of it is cooked in the kitchens on site. His ideas on sustainability is not made explicitly visible in the canteen, though he sources some of his food regionally, and uses organic and seasonal produce. One does encounter the presence of biodegradable cups and bowls in the canteen.

Habits, expectations and tastes might be challenged via the 'food experience strategy', where changes are being made in canteen food provisioning, such as replacing food products and

*Figure 6.1. Albert Kooy's catering concept services different kinds of freshly prepared foods, cooked and served by students, at different counters, e.g. 'Out of Africa', 'It's only bread', 'Made in Europe'.*

introducing others (e.g. organic food), changing recipes and menus (e.g. meals based on vegetables instead of meat), or making food provisioning more seasonal, etc. Albert Kooy mentioned consumers getting used to food that tasted different. Examples mentioned from his canteen are the bread that is made using more artisan methods and foods that do not contain any additives (like taste enhancers). Kooy indicated that his customers gradually have 'gotten used to' such changes over the years.

A case which illustrates the interaction between changes made in canteen food provisioning and the consumer more precisely is presented by the case below. Here canteen expectations and routines were challenged in such a way, as proved successful in the long run. Lars Charas was a cook hired by a Dutch research institution (RIVM) to improve on the quality and sustainability of their catering service (2005-2007). He was consciously aware that such an undertaking also meant challenging food expectations, habits and routines. He introduced a more varied and flexible menu, serving different kinds of meals, that is, more vegetarian, organic and seasonal dishes, as well as making meals with leftovers (Box 6.1). This was done gradually, over a period of time. He named creating a higher culinary quality and daily personal contact with his customers as

**Box 6.1. Creating a new canteen experience: Lars Charas' transformation of the canteen at RIVM.**

Within the RIVM canteen (catering for around 400 people a day) more sustainable provisioning was realised by making menu changes, i.e. introducing different recipes and ingredients. Charas was hired by the catering company in charge (large contract-caterer hired by RIVM) and given a lot of freedom by his catering manager to use his cooking and sourcing skills to employ more creative sourcing and cooking methods. Charas' main strategy involved making meals from fresh produce, making more use of seasonal and/or organic fruits and vegetables, and serving more vegetarian dishes. He sourced these from a local fruit and vegetable traders. Sticking to the fruits and vegetables of the season helped saving money. The local fruit and vegetable trader delivered produce pre-chopped, which saved time in the kitchen, making working with fresh produce easier. The cook used his own recipes to make special seasonal dishes and gave customers extra treats by baking cakes and making nice desserts. In addition, food wastage was minimised by using leftovers in a different dish the next day.
As for the communication strategy, Charas made sure that people knew him as the canteen's cook, showing is face regularly in the canteen and making it possible for customers to give feedback face-to-face. The cook described his take on sustainable canteen provisioning as follows: 'Sustainability should not just be present in an implicit way, only clear to those who are working with it or interested in it. It should be made explicit to users; one should be able to see it, smell it and taste it. Those are the preconditions to make sustainable catering successful.' Part of his method was to create an enjoyable, positive food experience and introduce an element of surprise. By introducing a greater variety of dishes and foods he believed customers would visit the canteen with more curiosity ('what's for lunch today?'), and be less likely to follow their old routine. Business lunches/dinners could be held at within the institute itself, since the food quality was high enough. Their ability to receive guests themselves made for good PR.

the main features of his strategy. His approach was to make the canteen food more tasty, makes changes in food sourcing and food preparation, but also to influence routines and conceptions about canteen food. He would make it a habit to ask customers if they had enjoyed their meal, and he gave explanations about the food and his approach to food. He also worked on making the canteen food service more interesting, adding an element of surprise: '... people's ideas are shaped by their routines. Take people out of their routine and you will have their attention for something new' (interview Lars Charas). Overall, a more varied and creative canteen menu was created, and with it end-users got used to this manner of catering, were 'seduced' by good tasting food, and better quality service.

When we speak about changing practices of food consumption, as we have seen in this and the previous chapter, the food practices of canteen consumers are not the only practices of interest for sustainable transition. The practices, that is, the habits, routines, concerns and 'portfolios' (knowledge and skills) of catering providers are equally important. In Chapter 5 we already discussed that there is a body of knowledge and skill needed for catering providers to be able to realise sustainable canteen transitions. This concerns knowledge and skill from sustainable procurement to what is happening within the kitchen. Then there is the 'canteen food culture' which is not only influenced by the customers, the end-users and organisations which hire caterers, but equally so by the caterers, the food professionals operating within the catering sector. These have ideas about what constitutes good catering, food quality, about what the role of the caterer is in sustainable development within catering ('follower' or 'innovator' role), about food wastage versus plastic waste (e.g. the use of monopackaging), about the extent to which food sourcing and food preparation should be standardised.

## 6.8 Discussion

### *6.8.1 The canteen lunch practice*

Participants in the end-user focus group described canteen food choices/practices in relation to work and study routine, and reasoned food choices in terms of 'what food looked good', 'where the shortest queue is', and perceptions of what was good value for money. In that sense, these descriptions from participating end-users are not so different from the kinds of findings found in other studies dedicated to understanding food practices (Carrigan *et al.*, 2006; Halkier, 2009a; Jabs *et al.*, 2007; Spaargaren *et al.*, 2013). These studies reveal how a person's reasoning about their food practices entail descriptions of how they manage and experience food intake in their everyday lives. Momentary enjoyment and 'practical' considerations may veer from, and even take over certain '(food) morals' or principles. Making a sustainable choice may be the intention, but everyday food consumption practice involves weighing up of a range of different considerations (see also results of focus groups by Spaargaren *et al.* (2013) on canteen lunch practice). Generally speaking, the employees and students who participated in the focus group expressed their support for a sustainably operating canteen and one which serves enough sustainably produced foods. However, the canteen context was considered different from their 'home-context'. Some expressed that they felt they have little influence on what is served in the canteen. Others expressed how issues of time and scheduling, tastes and fancies or other considerations play a role in their daily canteen food choice. Generally, one might conclude the practicalities of work and study schedules puts 'canteen consumers' in a position where they have limited incentive to get involved in catering.

Furthermore, judgements about the quality of the university's canteen food services, the food on offer there (taste, price) and the amount of sustainable alternatives available, were often related in terms of perceptions about value for money and/or past experiences with canteen food. Thus, opinions about catering are to some degree affected by past experiences with similar services. These experiences have built a frame of reference on food quality, price, taste, service quality, etc.

### *6.8.2 Encouraging sustainable food practices on the canteen floor*

Transitions toward more sustainability within catering services might be aided by realising a better 'fit' between food provisioning and the consumer. Facility experts have expressed the need for catering services within work-place and higher-education to be more focussed on the end-user. In some cases this might be realised by involving end-users more structurally in decision making on food and canteen procurement. Other options lie in gaining a better understanding of end-users practices, concerns and wishes, and/or framing food procurement in terms of certain organisations values and/or goals. In this way canteen food procurement can become more reflective of shared values within an organisation. Indeed support for changes toward more sustainable catering might be generated through a more profound discussion amongst the members of an organisation on the underlying assumptions about canteen food quality and sustainability issues, thus creating a feeling of shared responsibility for why certain procurement decisions are made.

All caterers interviewed for this research focussed on combining sustainability issues to some degree with the art of good food and good catering hospitality. In that sense they approach sustainability using the 'food experience strategy', and they combine sustainability with goals of maintaining a healthy and successful business (i.e. people, planet and profit). Here the focus is less on information provisioning and/or creating dialogue or debate on sustainability with consumers, although this may play a role, and more on developing the food service itself, and the locale where food consumption takes place. The basic tenant of this strategy is to connect with end-users through taste and enjoyment. Caterers are concentrating on the taste of the food they serve, its content and the way it is prepared (recipes and cooking), food (service) presentation and the way they communicate to their customers. Sustainability is an integral part of the canteen concept, but it is in fact not the main 'selling point'. The latter is gained mainly from creating quality within the service, value added and distinction in the service they provide. The 'food experience' strategy seems to contain the promise of being able to change every day canteen food consumption practices to some degree. Food habits and routines can be challenged through changing the canteen setting and the kind of food served there. End-users can be introduced to different kinds of foods, and meals, different tastes and contents. They can also get used to eating according to the seasons.

Whereas the practices of end-users was very much the starting point for this research, the study on catering has revealed how important the practices of canteen providers are. Making sustainable provisioning a success rests for a large part on the skills, knowledge, experience and creativity of providers. We have also seen that caterers operate in a certain context (business relationships, contracts, supply chain, etc.), with conceptions about what they consider to be 'good food' and 'good catering service'. The theoretical implications of this will be discussed in the following chapter.

# Chapter 7. Conclusions

## 7.1 Introduction

This study applied a practice approach in a sociological study of food consumption with the goal to provide further insights into the limitations and opportunities for sustainable development. In this way it hopes to provide an alternative vista to the more psychological-based perspectives currently applied to much of the research on sustainable consumption in the Netherlands, i.e. perspectives which focus on changing consumption behaviour through influencing values and attitudes (Reinders *et al.*, 2009). Focussing the attention on food practices has proven helpful in revealing the complexities of 'greening' ordinary, everyday consumption (Connolly & Prothero, 2008; Connors *et al.*, 2001; Halkier, 1999, 2001; Krom, 2008; Macnaghten, 2003; Maubach *et al.*, 2009; Wilk, 2006). Here, the consumer is not just a 'chooser' of products and services, but a practitioner managing everyday food practices. Consumers eat and buy food in different social contexts and settings, with specific food infrastructures and systems of provisioning servicing these 'sites of consumption'.

Using the social practice approach (SPA) (Spaargaren *et al.*, 2007a) allows food consumption to be seen as something which is socially and physically 'situated'. It offers a systematic way to consider what is happening at the locales of food consumption, at the interfaces or interactions between consumers and providers, when actors meet and incorporate structures. In this context, the consumer is a practitioner who partakes in a certain context or infrastructure of consumption, who is more/less equipped or more/less inclined to 'consume sustainably'. In this way, 'consumption' is not an expression of a subjective value structure but of a social interaction, with a specific setting and history. Such a theoretical approach offers interesting perspectives on sustainable transitions in everyday consumption. For example, Verbeek (2009) has provided insights into the role of specific travelling portfolios, i.e. bundles of (past) experience and practice-specific knowledge and routines in sustainable tourism consumption. This study made clear that different groups of consumers have different practice portfolios ('practitioners' experience/knowledge/skill) for specific types of environmentally friendly tourism mobility, with their different transport and tourism infrastructures and arrangements (p. 255). In a similar way, the present study identifies (groups of) consumers, with different portfolios pertaining to sustainable food behaviours, acting within and through a range of food contexts or sites while being more or less supported or constrained by the 'infrastructures' of food provisioning to consume sustainably.

The research for this study was carried out using a literature review, a quantitative study (survey) on the sustainable food practice of Dutch consumers and a qualitative study on the social practice of eating food in work-place canteens. The survey study shows to what extent Dutch consumers make use of different 'sustainable food alternatives' (e.g. 'product alternatives' like organic food and Fair Trade products, and 'practical alternatives' like eating less meat, cutting down on food wastage). It also shows to what extent consumers with different shopping practices differ in their use of these alternatives (thus making for varying portfolios), their opinions of these alternatives and their views on the role of the consumer in sustainable consumption. The catering study provides insight into an area of food consumption which is often overlooked; out-of-home food consumption in work and study-related contexts. It gives insight into how catering providers can realise sustainable food provisioning in the work-place canteen. It demonstrates the potential for sustainable development of everyday food consumption within and through catering services operating in the private and public sector, and the role of food professionals in this. The following sections will provide further discussion of the primary research questions of this study.

## 7.2 How to approach 'the consumer'

What insights does a practice based approach generate with regard to the development of more sustainable food consumption?

The dominant view on the sustainable development of consumption suggests that sustainable consumption is to be realised through behavioural changes originating from the individual consumer. In line with this, scientists and policymakers alike spend much effort to uncover the factors influencing food choice, and context is seen as just one of them. However, as Shove (2010) points out, a large conceptual omission in thinking about sustainable transitions within society is reducing context (infrastructure, social interactions, time constraints, etc.) to an external causal factor, one in line with other factors that may influence behaviour (e.g. habit, attitude, value structures, etc.). The problem here is that context is never external to behaviour, it is implied in it. A practice-based approach serves to insert behaviour into context. It directs attention to questions concerning the interactions between actors and between action and structures. It opens up the possibility to ask questions such as: How have these unsustainable food arrangements come about? How are they organised, who has which hand in organising them? It also encourages the search for the opportunities, the potentials which lie within contextual interactions to positively influence consumption patterns.

Chapter 3 endeavoured to take a bird's-eye view of today's food consumption practices. Its objective was to describe some of the most salient features of daily eating and food shopping practices. With an orientation on practices in mind this meant concentrating on everyday eating and shopping for food in different situations and contexts. The three main characteristics discussed were the need for convenience in both food preparation and shopping, the use of food service(s) in varying forms and contexts outside the home and the diverse and fragmented nature of daily food consumption practices, as well as the marginal direct involvement of consumers in the preparation of their daily meal. These features impact the way in which we view the role of the consumer in sustainable development. In principle consumers certainly carry part of the responsibility for supporting unsustainable agro-food systems, and there is a degree of empowerment which consumers can have and act on in 'boycotting' certain products, choosing alternative food products and services and participating in and making use of alternative food networks. Micheletti (2003) points out how these alternative choices are political and even powerful. With consumers acting within the market in such a way, leverage might be found to exert certain changes. However, an overemphasis or overestimation of either the responsibility or the power of consumers discounts their varying roles and positions (Jacobsen & Dulsrud, 2007) and can be used as an excuse for inaction by other actors (business, government) (Warde, 2013a).

Consumers make use of different food providers in different contexts and the involvement of and access to sustainable products and services in these contexts is not constant, with consumers having different needs and considerations and in many instances very little influence over the content of their meals. Thus, what is conceptualised and described as 'individual' consumption/consumer food choice is at best often a 'shared consumption choice', i.e. involving a chain of decision making by various actors, involving specific food infrastructures or systems of provisioning. In this respect 'food consumption' not only concerns people who shop in supermarkets, or eat a meal in a restaurant, etc., consumption also incorporates organisations, buyers, food designers, nutritionists, (facility) managers, cooks, caterers, etc. The challenge is to find 'leverage' in the contexts of food consumption; through food infrastructures and locales of consumption, providers and the interactions between providers and consumers, and in understanding consumers' practices.

Chapter 4 gives us some insights into the role which Dutch consumers see for themselves in terms of their co-responsibility for sustainable consumption, their opinions on a range of different ways to 'behave' sustainably (the use of various sustainable alternatives) and their use of these alternatives. Although many Dutch consumers think consumers are partly responsible for environmental problems caused by food consumption, many are undecided as to the active role of consumers in terms of the use of sustainable food alternatives. It seems that Dutch consumers differ more in their sustainable portfolio (food sustainability knowledge and experience with certain sustainable food alternatives) than in their awareness about the link between the environment and daily food consumption. Consumers who shop at alternative shops (farmers markets, organic shops) do not appear to be 'more aware' of the link between the environment and food than consumers who only shop at supermarkets. They do however make more use of 'product alternatives' such as organic and Fair Trade foods. The difference in experience with sustainable food alternatives is seen most in the use of certain products, like organic and Fair Trade labelled foods, not in the use of 'practical alternatives' like cutting down on food wastage. If nothing else, these results question typical preconceptions about what defines an 'ethical' or 'green' consumer. Similarly, Verbeek (2009) found that different clusters of consumers had different practical experiences with environmentally friendly forms of travelling on holiday, but were not very different in their abstract concern for the environment. Her study shows how consumers are not so easily (or should not so easily be) categorised as 'more' or 'less' (or 'light' and 'dark') 'green' in terms of their generic environmental attitude. Instead, consumers have a certain set of practice-based knowledge and experience with which future choices within mobility practices are made and which 'give permanence to tourism mobility routines' (p. 243).

The catering study (Chapters 5 and 6) sheds light on food consumption in the context of public/private food procurement (e.g. for canteen food in the work place). It illuminates food consumption in the relatively important and overlooked daily food practice of 'eating at work or in a study-related setting', and broadens our horizon on the consumer. Here we see that consumers are moving within a different context compared to home-based/related practices (like grocery shopping). Consumers 'at work' are relatively uninvolved and 'out-of-touch' with the food provisioning organised by the organisation where they work/study. Often catering services within the work place are arranged in a relatively top-down manner and there is inadequate feedback between end-users and catering providers. However, there is great potential for influencing or orchestrating more sustainable food consumption practices in canteens.

Firstly, there are various ways to make consumers more involved in catering provisioning and/or to make sustainable food procurement a more legitimised and 'shared' undertaking within an organisation. The shared, collective nature of food consumption within organisations could be used as a starting point for approaching sustainable food procurement within organisations. The idea is to (re)-connect the end-user with catering procurement decisions, making them 'co-producers' in such a way as to function as a positive or stimulating factor for sustainable food consumption.

Secondly and maybe more important, innovative caterers showed how changes could be made in menus and recipes in order to realise more sustainability (e.g. more vegetables and less meat, more seasonal food, more organic food). These changes were made acceptable to consumers through good tasting, freshly prepared food in combination will well-designed food services. The focus group and the interviews showed that innovative caterers prefer concentrating on their 'core business', creating distinctive food experiences within the canteen context/practice, thus using their culinary knowledge and entrepreneurial skills, instead of 'educating' consumers through more rationalised information and labelling systems.[60] Providing information might play a role in creating awareness and appreciation about what the caterers are doing, but in the canteens visited for this research the focus was primarily on food enjoyment.

The catering study illustrates the role of practical and context specific experience(s) in shaping expectations and norms related to food and in this case the catering service context. Employees and students participating in the focus group discussed their opinions about the sustainable provisioning in their canteen in terms of value for money and experiences in other canteens. They also referred to the momentary considerations that play a role in their lunch practice, for example with regard to which food looks good and which queue is shortest. This finding reiterates the idea that opinions and frames of reference relating to food are shaped via practical routines and experiences within specific contexts, in interaction with specific food infrastructures, i.e. locales and systems of provisioning, and food providers (cooks, catering managers). Within such dynamics there is the opportunity for contextualised learning, for creating practise related food experiences which create positive frames of reference about for instance organic foods, vegetarian meals, seasonal vegetables, etc. Here lies great potential for experimenting with 'new ways to eat', not so much via abstract channels of communication, but via concrete positive associations, and building values related to food and agricultural quality within society.

By way of conclusion, we would argue that a practice based approach offers insights which go beyond the realm of abstract, individual and/or structural conditions alone. In the survey study this was especially apparent when the findings showed that consumers with different shopping practices tend to differ more in their use of certain product alternatives than in typical ideas associated with awareness of sustainability issues. These results suggest that sustainable consumption behaviour is connected to what is going on in the realm of social practices of food consumption and the interaction with infrastructures of consumption (locales and systems of provisioning). As in the study of Verbeek (2009), the survey suggests a blurring of the distinction between topics and issues that are commonly used to construct 'green' and 'non-green' or less-green consumers.

[60] Sustainability issues might be mentioned in websites and mission statements, and in communications with their contract-lender/employer, it was rarely communicated through certain design features or present in the form of signs, posters or information leaflets for example.

Instead, a more contextual specific mixture of 'thinking and doing' is shown. The catering study demonstrated the thoroughly contextual nature of food consumption practices, in this case in the work place. Both studies (catering case and survey study) suggest that a diversity of factors and dynamics come together and work together to produce a particular, context specific contribution to sustainable development in the food sector. The catering study particularly illustrates how 'sustainable consumption' is orchestrated, experienced and learned in group- and site-specific ways. By looking into actors in the systems of food provision, the important and influential roles of food professionals in enabling sustainable food consumption of ordinary consumers became clear, and the practices of food professionals, i.e. their knowledge and ideas about food, their creativity and competence, were recognised and given proper analytical weight. Finally, a more practice based approach allows for a better understanding of the opportunities for change and the leverage points for transitions into more sustainable food consumption in the future. The survey study and the catering study present a number of suggestions for transition management and for strategic action as discussed in the next paragraphs.

## 7.3 Strategies for sustainable food consumption

What kind of strategies for the transition to more sustainable food consumption practices can be derived from studying food consumption using a practice-based approach?

### *7.3.1 More attention on wastefulness*

The survey study (Chapter 4) revealed that cutting down on food packaging and food wastage, keeping to food seasonality and buying foods of local/regional origin appear to be the most attractive and firmly embedded within the everyday food consumption routines of Dutch consumers. This did not vary much between consumers who shopped at alternatives shops, or consumers who only visited supermarkets for their grocery shopping. Within the focus group on sustainable catering the topic of waste also came up spontaneously amongst the end-user participants. Here the (mono)packaging of food products within the canteen was discussed and was considered an environmentally unfriendly feature of canteens (Chapter 6). It seems that there is a great affinity with the concept of wastefulness under Dutch consumers, while at the same time there seems to be ample space for action, both within shopping and within canteen environments. Considering that food wastage, packaging and seasonality are important to such a large proportion of Dutch consumers, food providers might benefit from incorporating them more actively into their sustainable development strategies. Until now many food retailers and service providers who offer more sustainable products and services profile themselves in terms of naturalness, taste, quality, organic farming, etc. in other words on features concerning the foods itself, and its origins. Waste and wastefulness are not topics that are explicitly seen in concepts, while this directly touches upon the materiality and the experience of food consumption practices in both shopping and canteen eating settings.

### 7.3.2 The food experience strategy

Canteens represent locales for the generation of positive food experiences with different foods and dishes (Lassen *et al.*, 2004). The canteen represents a setting in which sustainable food practices can be contextualised and (re) framed, a process in which food professionals play a vital role. Indeed, amongst most of the caterers interviewed for this study, or those who participated in the focus group, there seems a broad consensus that 'food experience' is the way to realise sustainable canteen food provisioning and consumption. The cases discussed in the catering study reveal how caterers were able to successfully introduce more sustainable food provisioning by targeting the food itself: the taste of the food, its content and appearance, food presentation and the relationship between caterer and end-user. The inherent advantages and benefits of food become tactile; consumers can 'experience' the food for themselves. We also saw how changing to more sustainable food provisioning in canteens involved a 'soft' challenging of both consumers' and providers' practices. Canteen providers have to find/learn new skills and knowledge to adapt food provisioning and preparation, new conceptions of what constitutes good canteen food and good service. In turn, end-consumers were sometimes confronted with a different approach to food, with different tasting foods to which they had to get used to, or the switch from standardised menus to more dynamic (e.g. seasonal), fluctuating menus.

Attention for culinary content can be used as a strategy for improving on both health and sustainability. For example, creating tastier, healthier meals can be seen as a strategy for sustainable provisioning, for example, a menu with a higher fresh vegetable content. Even amongst nutrition scientists culinary quality and food enjoyment seems to be getting more attention. For example, in the Netherlands food nutrition researchers looking into improving diets of the elderly have recognised that the quality of food services within elderly care needs to be improved (see project 'de Genietende Groene Tafel' and 'Genieten aan Tafel' by Wageningen University and Research Centre) (Brok & Gorselink, 2010). In the Netherlands many people in elderly care homes are underfed or undernourished. Researchers looking into this problem have found that food enjoyment, i.e. good tasting food, nicely laid tables and pleasant contact between catering personal and customers, are needed to improve the dietary health of elderly people. This requires investment which, so it is argued, will be offset by healthier individuals who may then for example require less medicine.

The food experience strategy is so interesting because it touches the heart of notions of food consumption practices. The understandings and meanings which guide people's thinking about food may be connected to certain values, but they are rooted in direct experiences of practicing shopping, cooking, etc. (Aubrun *et al.*, 2005; Spaargaren *et al.*, 2007a). In the catering study we saw how people's expectations and attitudes in relation to canteen food were influenced by (prior) experience(s). Creating positive experiences with sustainable food provisioning and thus also building a 'portfolio' which represents and/or enables a 'sustainable food lifestyle' is something which involves working on concrete interactive infrastructures of food provisioning, distribution and consumption. That is, the availability of sustainable, good quality food products, good food infrastructure in various areas of society, from hospitals, to schools to offices. Building on interactive infrastructure which stimulate quality of life, quality of bodies and minds. This is something which always begins with changing institutions and organisations and fostering and encouraging innovation, not with changing consumers.

Creating a better food infrastructure includes competence building in this area and investing in physical infrastructure, i.e. kitchen and canteen facilities. Realising more sustainable canteen food provisioning via the food experience strategy is aided by a competent cook but also a fully functioning kitchen. A caterer who has a fully functioning kitchen to his/her disposal will be able to 'create' his/her own food and be more creative in finding solutions with respect to food sourcing and preparation. If the kitchen is not furnished for cooking there is no need for a great deal of cooking expertise and the caterer will be more dependent on suppliers which supply pre-prepared meals and meal components. In other words, caterers will be more dependent on the developments within the food chain and industry to provide them with pre-packaged, pre-prepared meals and meal-components, that are sustainably produced and/or sourced. They will be dependent on suppliers, wholesalers, the food industry to provide them with hem with the kinds of products with which they might add value to their services, in other words, products which have a certain level of quality, or have some added advantage or attribute which the caterer might be used to market within the canteen.

### 7.3.3 Recognising the role of food professionals

It is the actors that are in charge of public/private (company) food procurement and the cooking, handling and presenting of food who have an important part in creating sustainable food consumption practices. They are in the unique position to influence people's ideas about food, their habits, tastes and diets. In other words, they are in a position to influence food culture. In this research the focus has been primarily on the caterers; the food professionals who design the locales of food consumption and organise the food provisioning of these locales. Making changes in sourcing, food preparation and menu's in order to realise more sustainable food provisioning requires trained food professionals. The latter require knowledge about food sustainability and culinary knowledge, but also cooking and canteen management skills.

The role of facility managers and other actors within organisation to make decision on food procurement and catering have been left largely outside of scope of this study. Sustainable food procurement requires knowledge about procurement regulations and guidelines (EU and national) but also a dose of creativity in order to realise affordable sustainable provisioning fitting the wishes of an organisation (e.g. to realise more locally sourced food provisioning). It also requires insight into the effects which contracts have on the sustainable development of the canteen, which types of contracts are stifling to sustainable development and which contracts promote it. What kinds of relationships between suppliers and caterers, caterers and contract-lenders, and canteen providers and end-users promote sustainable development and which do not. Sustainable development should be seen as a process, i.e. something which needs to 'develop' and move forwards, not something which is static. Some caterers (interviews and focus groups) pointed out that some facility managers and buyers of organisations focus too much on functionality and cost-efficiency, a kind of culture which appears stifling to the creativity and skill which a caterer might need to create 'food experience' in a canteen, both with regard to the food as such as with regard to the 'staging' of it in canteen settings. Here too, we see that the culinary and service expertise of food professionals is maybe not employed to its full potential.

Food professionals might be interesting partners for policymakers because of their central role of influence on food consumption practices. Here lies a pool of untapped (potential) creativity and (potential) competence. Setting targets and guidelines for the procurement of organic food, Fair Trade food, etc. as done in the national guidelines for sustainable procurement policy for catering within the public sector, is one thing. Harbouring and stimulating a creative and innovative food profession is another. Here, the focus is not on individualised targets, of individualised products (e.g. 40% of food procured by provinces and municipalities should be organic) (VROM, 2010), but on creating a sustainable 'meal', a sustainable lunch, and access to these meals. The latter can in effect only be realised by creating an infrastructure which can provide in this. For sustainable development in this area of food consumption, professionals are needed who are capable to find sustainable solutions based on their craft. Not all cooks and caters are innovators and creative spirits. It might be worth wile to take a closer look at the education of food designers, cooks and facility/catering managers. The educational programmes of these professions (e.g. hotel management, facility management studies) need to include training on sustainable food provisioning and management.

## 7.4 Reflections on methodology and recommendations for further study

### *7.4.1 The survey study*

The survey study provides a more practice-oriented look at sustainable food consumption in the sense that it highlights how Dutch consumers see their own active role in sustainable food consumption, the kinds of sustainable food alternatives they consider attractive and their engagement in the use of a range of different sustainable food alternatives in their daily life. It also provides some insight into the extent to which this engagement differs between consumers with different shopping practices (expressed in terms of type of shops visited). The survey also indicated that Dutch consumers consider sustainability considerably less during visits to canteens, cafes and restaurants than during grocery-shopping. Had the survey been set up in such a way as to refer questions to one or more specific practices, including for example in explicit reference to food consumption within the out-of-home sphere, this might have more clearly shown how different practices relate to different sets of considerations and portfolios. Further development of the survey in this way might show how different kinds of alternatives are considered attractive in different practices, and how practical knowledge and experience differs between different practices (e.g. shopping for food, eating out, etc.).

The survey only provided a general understanding of Dutch consumers' sustainable consumption portfolio, that is, experience with/use of sustainable food alternatives and practice-related ('action-related') knowledge and skills for engaging in sustainable food consumption in daily life. It also gave insight to the extent to which these differ between consumers with different shopping practices. However, there are more 'portfolio aspects' which might be considered, and which might play a role in the level of sustainable consumption which can be attained. For instance, if we think in terms of the everyday shopping and cooking practices one might consider consumers' 'green shopping, cooking and food management skills'. For example, do people 'know' (in the sense of a practical knowing, not so much just in terms of an abstract knowledge) where to find sustainable food products (vegetables, meat, fish)? Do they have certain 'green' food management skills and cooking skills? For example, do they know how to cook with less meat and/or know how to substitute meat with something else (e.g. vegetables, legumes)? Are they able to realise goals of taste, convenience or cost in combination with sustainability? How do people with different budgets engage in sustainable food consumption? In other words, investigating Dutch consumers potential portfolio with regard to making their daily food consumption practices more sustainable, something which provides a truer picture of consumers as 'sustainable food practitioners'.

### 7.4.2 The catering study

The catering study revealed that not only consumers' food practices are of interest, but equally interesting and important to consider are the practices of catering providers. Caterers themselves develop (company) strategies which imply certain conceptions about what constitutes good catering services, how to realise sustainable provisioning, etc. The extent to which they can 'green' their business depends upon their portfolio, the professional skills and knowledge needed to provide a sustainable food service. It might be interesting to see how this idea of provider practices can be used to move away from the idea that more sustainable food consumption is primarily or exclusively attached to practices enacted by consumers. One might conclude that the practices of catering providers come together with end-users practices, the two influence each other[61], and the dynamic between these practices result in certain consumption patterns.

In the catering study we chose not to go into the role of the employer/contract lender (the organisation/institution where the canteen is situated). Further study could be dedicated to the subject of contract setting for sustainable food procurement, and the roles and capacities of different actors here. Programmes in the Netherlands are already focussing on encouraging sustainable procurement within organisations of the public sector (municipalities and province-offices). Here there is the idea of 'setting a good example' and providing an impulse for sustainable production and consumption. However, how can organisations in other sectors be moved to provide more sustainable food services? How can sustainable procurement really be encouraged in the private sector without the intervention of government? It would be good to re-examine the effectiveness of the government and of governance (Dutch and/or EU) in encouraging sustainable food procurement of organisations. In the past, governance on sustainable food consumption and sustainable procurement has run along the lines of information provisioning and building public-private partnerships. Food consumption policy is characterised by the focus on improving communication and marketing, assisting and facilitating public-private partnerships with major players within food retail and services. A critical analysis of this path is needed, but also an analysis of how governance can support/facilitate sustainable development initiatives within the public and private sphere; where providers and consumers are organising themselves to create sustainable food networks. Furthermore, how does the Dutch government related to the public plate today? How can executive managers and board of directors of institutions be encouraged to

[61] The latter is seen in the standardisation of canteen food services, and the standardisation of consumption practice to match.

take up goals on sustainable and healthy food practice?[62] Would a good public plate infrastructure not mean savings on an enormous societal health burden? Is healthy and sustainable eating not inexplicably linked as a policy goal?

### *7.4.3 Food infrastructure and the public plate*

The 'public plate' refers to the food infrastructure which provides food, or meals, outside the context of one's own kitchen and cooking skills. This is the food supply in offices and places where public services are being provided like colleges, universities, schools, hospitals, day-care centres, and elderly homes. Improving the food infrastructures of these places presents the opportunity to combine various social, economic and environmental policy goals. Issues like children's and youngsters diets, education about food and cooking, improving the diets of elderly people in care-homes and the access to healthy food for elderly people at home, as well as sustainable development of regional and local agro-food business and networks. The proportion of people who would benefit from a good food service infrastructure is increasing (Bijl *et al.*, 2011). This especially includes children, young adults and the elderly. Over the years there has been an increase in one-person households and single-parent families, especially in urban areas (Bijl *et al.*, 2011). Though there has not been much research done on the state of Dutch cooking practices, judging from the range of pre-prepared and pre-cooked meals available in supermarkets today, many Dutch consumers are forgoing cooking 'from scratch' and making use of convenient food solutions. However, the impact of fresh foods (freshly prepared meals) remains paramount for a healthy diet. These factors would imply that there is room for growth for (better) food services (Hermens, 2012). Such a study area also brings with it some governance questions. First and foremost, how can a more 'social' policy on sustainable food infrastructure of the public plate be realised? Especially in light of the dominant paradigm of government keeping to the side-lines when it comes to influencing people's diets. For instance, the problem with improving the food infrastructure within Dutch primary and secondary education is that there is no public structure of responsibility to provide the daily meal for children and young people. This is primarily up to the parents and to some extent the schools themselves. This lack in structural responsibility and service mentality as to provide food infrastructure is seen in the absence of kitchen and canteen facilities, the presence of vending machines offering primary snacks and drinks high on refined sugars and the short time reserved for lunch. A lack of quality food services within care-homes, hospitals, schools and colleges, represents a kind of structural poverty of diet. However, what would be the costs saved on public spending on human health and environmental degradation if diets improved structurally?

[62] For instance sustainable policy is only considered by 4% of the board of directors of Dutch care/elderly-homes (Hermens, 2012).

# References


Adrichem, L., van (2009). The dream canteen. Verslag van een bezoek aan de Salone del Gusto, Turijn 2008. Utrecht, the Netherlands: InnovatieNetwerk.

Aubrun, A., Brown, A., & Grady, J. (2005). *'Not while I'm eating'. Perceptions of the U.S. food system: what and how Americans think about their food.* East Battle Creek, MI, USA: W.K. Kellogg Foundation.

Bakker, J. (2007). Bio-monitor 2007 Jaarrapport. Biologica and task force marktontwikkeling biologische landbouw. Available at: http://edepot.wur.nl/136772.

Baltussen, W.H.M., Wertheim-Heck, S.C.O., Bunte, F.H.J., Tacken, G.M.L., Galen, M.A., van, Bakker, J.H., & Winter, M.A., de (2006). Een biologisch prijsexperiment; grenzen in zicht? Den Haag, the Netherlands: LEI.

Barnes, B. (2001). Practice as collective action. In: T.R. Schatzki, K. Knorr Cetina & E. Savigny, von (Eds.), *The practice turn in contemporary theory.* London, UK: Routledge.

Barr, S., & Gilg, A. (2007). A conceptual framework for understanding and analysing attitudes towards environmental behaviour. *Geographical Annual, 89B*(4), 361-379.

Bartels, J., Onwezen, M.C., Ronteltap, A., Fischer, A.R.H., Kole, A.P.W., Veggel, R.J.F.M., van, & Meeusen, M.J.G.R. (2009). Eten van waarde; peiling consument en voedsel. Den Haag, the Netherlands: LEI.

Bava, C.M., Jaeger, S.R., & Park, J. (2008). Constraints upon food provisioning practices in 'busy' women's lives: trade-offs which demand convenience. *Appetite, 50*(2-3), 486-498.

Beardsworth, A., & Keil, T. (1997). *Sociology on the menu: an invitation to the study of food and society.* London, UK: Taylor & Francis Group.

Beekman, V., Kornelis, M., Heijden, C., van der, Aramyan, L., Vollebregt, M., & Herpen, E. (2007). In gesprek over voedselkwaliteit; Het ministerie van LNV midden in de samenleving. Den Haag, the Netherlands: LEI.

Benschop, I. (2008). Biologische catering bij kleine organisaties. Provinciale Milieufederaties. Available at: http://tinyurl.com/nfbsr8a.

Berends, W. (2004). Van grond tot mond: transparantie in de voedselmarkt. Utrecht, the Netherlands: Stichting Natuur en Milieu.

BHeC (2003). Kompass voor beleid voor 2004. Zoetermeer, the Netherlands: Bedrijfschap Horeca en Catering.

BHeC (2004). Contractcatering in Nederland. Zoetermeer, the Netherlands: Bedrijfschap Horeca en Catering.

BHeC (2005). Marktpotentie van bedrijfscatering in Nederland. Zoetermeer, the Netherlands: Bedrijfschap Horeca en Catering.

Bijl, R., Boelhouwer, J., Cloïn, M., & Pommer, E. (2011). De sociale staat van Nederland 2011 (Vol. scp-publicatie 2011-39, pp. 352). Den Haag, the Netherlands: Sociaal en Cultureel Planbureau.

Blake, J. (1999). Overcoming the 'value-action' gap in environmental policy: tensions between national policy and local experience. *Local Environment, 4*(3), 257-278.

Bloor, M., Frankland, J., Thomas, M., & Robson, K. (2001). *Focus groups in social research.* London, UK: Sage Publications.

Boer, J., de, Hoogland, C.T., & Boersema, J.J. (2007). Towards more sustainable food choices: value priorities and motivational orientations. *Food Quality and Preference, 18*(7), 985-996.

Bourdieu, P. (1977). *Outline of a theory of practice.* Cambridge, UK: Cambridge University Press.

Bouwman, L. (2009). *Personalized nutrition advice; an everyday-life perspective.* PhD thesis, Wageningen, the Netherlands: Wageningen University.

Bowman, S.A. (2006). A comparison of the socioeconomic characteristics, dietary practices, and health status of women food shoppers with different food price attitudes *Nutrition Research, 26*(7), 318-324.

Brok, P., den, & Gorselink, M. (2010). De genietende groene tafel. Wageningen, the Netherlands: Phliss and Wageningen UR Food & Biobased Research.

Brug, J., & Lenthe, F.J., van. (2005). *Environmental determinants and interventions for physical activity, nutrition and smoking: a review*. Den Haag, the Netherlands: ZonMW.

Brunso, K., & Grunert, K.G. (1995). Development and testing of a cross-culturally valid instrument: food-related lifestyle. *Advances in Consumer Research, 22*, 475-480.

Bryman, A. (1999). The Disneyization of society. *The Sociological Review*, 25-47.

Buckley, M., Cowan, C., & McCarthy, M. (2007). The convenience food market in Great Britain: convenience food lifestyle (CFL) segments. *Appetite, 49*(3), 600-617.

Buckley, M., Cowan, C., McCarthy, M., & O'Sullivan, C. (2005). The convenience consumer and food-related lifestyles in Great Britain. *Journal of Food Products Marketing, 11*(3), 3-25.

Bunte, F.H.J., Meeusen, M.J.G., Kole, A.P.W., Stijnen, D.A.J.M., Spiegel, M., van der, Bakker, J.H., & Kuiper, W.E. (2008). Eten van waarde. Voedselkwaliteit in Nederland. Wageningen, the Netherlands: Agrotechnology and Food Sciences Group, WUR.

Burgess, J., Harrison, C.M., & Filius, P. (1998). Environmental communication and the cultural politics of environmental citizenship. *Environment and Planning A, 30*(8), 1445-1460.

Burgess, S.M. (1992). Personal values and consumer research: a historical perspective. *Research in Marketing, 11*(35-79).

Carrigan, M., Szmigin, I., & Leek, S. (2006). Managing routine food choices in UK families: the role of convenience consumption. *Appetite, 47*(3), 372-383.

Carrington, M.J., Neville, B.A., & Whitwell, G.J. (2014). Lost in translation: exploring the ethical consumer intention-behavior gap. *Journal of Business Research, 67*(1), 2759-2767.

CBS (2007). Landelijke Jeugdmonitor: Rapportage 3[e] kwartaal 2007. Heerlen, the Netherlands: Centraal Bureau voor de Statistiek.

CGE&Y (2002). De foodserviceketen 'revolueert'. Succes dwing je af! Utrecht, the Netherlands: Cap Gemini Ernest and Young. Available at: http://tinyurl.com/popsckk.

Cheng, S.-L., Olsen, W., Southerton, D., & Warde, A. (2007). The changing practice of eating: evidence from UK time diaries, 1975 and 20001. *The British Journal of Sociology, 58*(1), 39-61.

Clarke, I., Hallsworth, A., Jackson, P., Kervenoael, R., de, Aguila, R.P., del, & Kirkup, M. (2006). Retail restructuring and consumer choice 1. Long-term local changes in consumer behaviour: Portsmouth, 1980-2002. *Environment and Planning A, 38*(1), 25-46.

Cohen, S., & Taylor, L. (1992). *Escape attempts: the theory and practice of resistance to everyday life* (2[nd] ed.). London, UK: Routledge.

Connolly, J., & Prothero, A. (2008). Green consumption: life-politics, risk and contradictions. *Journal of Consumer Culture, 8*(1), 117-145.

Connors, M., Bisogni, C.A., Sobal, J., & Devine, C.M. (2001). Managing values in personal food systems. *Appetite, 36*(3), 189-200.

Crawford, D., Ball, K., Mishra, G., Salmon, J., & Timperio, A. (2007). Which food-related behaviours are associated with healthier intakes of fruits and vegetables among women? *Public Health Nutrition, 10*(3), 256-265.

Damen, A. (2010a). Cateraar voldoet niet aan verwachting Universiteit. *Sense FM*.

Damen, A. (2010b). Onderzoek onder HBO's: een magere 7 voor restaurant. Cateraar te weinig ondernemer. *Sense FM*.

Damen, A., & Meer, E., van der (2010). Gast staat niet centraal. *Sense FM.*

Dagevos, J.C. (2004a). The proliferation of organic food consumption beyond the marketing mix. *Food Economics,* in press.

Dagevos, J.C. (2004b). Voedsel als uitdrukking van een levensstijl? Een sociologische benadering. *Ethische Perspectieven, 14,* 413-428.

Dagevos, J.C. (2005). Consumers as four-faced creatures. Looking at food consumption from the perspective of contemporary consumers. *Appetite, 45*(1), 32-39.

Dagevos, J.C., & Bakker, E., de (2008). Consumptie verplicht; Een kleine sociologie van consumeren tussen vreten en geweten. Den Haag, the Netherlands: LEI.

Dagevos, J.C., Herpen, E., van, & Kornelis, M. (2005). *Consumptiesamenleving en consumeren in de supermarkt.* Wageningen, the Netherlands: Wageningen Academic Publishers.

Dagevos, J.C., & Munnichs, G. (2007). De obesogene samenleving Maatschappelijke perspectieven op overgewicht. Den Haag, the Netherlands: LEI, Rathenau Instituut.

DEFRA (2006). Procuring the future: sustainable procurement action plan. London, UK: DEFRA.

Distrifood (2008, 11-10-2008). Klant laat keuze supermarkt afhangen van aanbiedingen en folder. *Distrifood Nieuwsblad nr. 41.*

Dixon, J., & Banwell, C. (2004). Re-embedding trust: unravelling the construction of modern diets. *Critical Public Health, 14*(2), 117-131.

Dogme2000 (2007). Dogme 2000 – A manual of a municipal environmental cooperation in progress. Available at: http://tinyurl.com/lbzkyor.

Drunen, M., van, Beukering, P., van, & Aiking, H. (2010). The true price of meat (Vol. Report W10/02aEN 4 May 2010). Amsterdam, the Netherlands: Institute for Environmental Studies, VU University.

EC (2004a). Directive 2004/18/EC of the European Parliament and of the Council of 31 March 2004 on the coordination of procedures for the award of public works contracts, public supply contracts and public service contracts. *Official Journal of the European Union, L 134*(3-4-2014), 114-240.

EC (2004b). Sustainable consumption and production in the European Union. Brussels, Belgium: European Commission.

EC (2005). Attitudes of consumers towards the welfare of farmed animals *Special Eurobarometer 229.* Brussels, Belgium: European Commission.

EC (2009). Contributing to Sustainable development: the role of fair trade and nongovernmental trade-related sustainability assurance schemes. Brussels, Belgium: Communication from the Commission to the Council of the European Parliament and the European Economic and Social Committee.

EC (2011). *Sociaal kopen gids voor de inachtneming van sociale overwegingen bij overheidsaanbestedingen.* Brussels, Belgium: European Commission.

Engstrom, R., & Carlsson-Kanyama, A. (2004). Food losses in food service institutions examples from Sweden. *Food Policy, 29*(3), 203-213.

Eriksen, T.H., & Nielsen, F.S. (2001). *A history of anthropology.* London, UK: Pluto Press.

Erkkilä, A.T., Sarkkinen, E.S., Lehto, S., Pyörälä, K., Uusitupa, M.I.J. (1999). Diet in relation to socioeconomic status in patients with coronary heart disease. *European Journal of Clinical Nutrition, 53*(8), 662-668.

Ernst, M. (2007). Biologische warme maaltijden in de Zeeuwse zorg: Evaluatie pilot 2006-2007. Middelburg, the Netherlands: Zeeuwse Milieufederatie (Provincie Zeeland, LNV).

Eurocities (2005). The CARPE guide to responsible procurement. Brussels, Belgium: Eurocities.

EZ (2012). Monitor duurzaam voedsel 2012 (pp. 56). Den Haag, the Netherlands: Ministerie van EZ.

Factcard (2004). Contractcatering a la carte 2004. Zoetermeer, the Netherlands: Bedrijfschap Horeca en Catering.

FERCO (2009). The contract catering market in Europe: 2006-2010, 25 countries. Brussels, Belgium: FERCO.

Fine, B., & Leopold, E. (1993). *The world of consumption*. London, UK: Routledge.

Fischler, C. (1980). Food habits, social change and the nature/culture dilemma. *Social Science Information, 19*(6), 937-953.

Fishbein, M., & Ajzen, I. (1975). *Belief, attitude, intention, and behavior: an introduction to theory and research*. Reading, MA, USA: Addison-Wesley.

FNLI (2008). Actieplan zout in levensmiddelen. Rijswijk, the Netherlands: Federatie Nederlandse Levensmiddelenindustrie. Available at: http://tinyurl.com/oswd72g.

Foodstep (2005). Het Nationale Cateringonderzoek 2005. Bennekom, the Netherlands: Foodstep. *Conclusions*

FSIN (2008). Quo Vadis?! Retail in 2007-2015. Available at: http://tinyurl.com/ov3fj9l.

FSIN (2009). Foodservice Markt 2009. Ede, the Netherlands: Food Service Instituut Nederland.

Gabriel, Y., & Lang, T. (1995). *The unmanageable consumer: contemporary consumption and its fragmentation*. London, UK: Sage.

Garnett, T., (2008). Cooking up a storm food, greenhouse gas emissions and our changing climate. Brighton, UK: Food Climate Research Network, Centre for Environmental Strategy, University of Surrey.

Geels, F. (2004). From sectoral systems of innovation to socio-technical systems. *Research Policy, 33*(6-7), 897-920.

Gezondheidsraad (2003). Overgewicht en obesitas. Den Haag, the Netherlands: Gezondheidsraad.

Gezondheidsraad (2006). Richtlijnen Goed Voeding 2006. Den Haag, the Netherlands: Gezondheidsraad.

Gfk (2008). Kant-en-klaar groeit ondanks ongezond imago. Available at: http://tinyurl.com/nncqyqb

Giddens, A. (1984). *The constitution of society*. Cambridge, UK: Polity Press.

Giddens, A. (1991). *Modernity and self-identity: self and society in the late modern age* Cambridge, UK: Polity Press.

GM&FNE (2007). Biologische catering bij (semi-) overheidsinstellingen benchmark 2006 (Final Report 2007). Arnhem, the Netherlands: Gelderse Milieufederatie and The Foundation for Nature and the Environment (GM&FNE)

Goffman, E. (1974). *Frame analysis: an essay on the organization of experience*. Lebanon, NH, USA: Northeastern University Press.

Grunert, S.C., & Juhl, H.J. (1995). Values, environmental attitudes and buying of organic foods. *Journal of Economic Psychology, 16*(1), 39-62.

Halkier, B. (1999). Consequences of the politicization of consumption: the example of environmentally friendly consumption practices. *Journal of Environmental Policy and Planning, 1*, 25-41.

Halkier, B. (2001). Consuming ambivalences: consumer handling of environmentally related risks in food. *Journal of Consumer Culture, 1*(2), 205-224.

Halkier, B. (2004). Consumption, risk, and civic engagement: citizens as risk-handlers. In M. Micheletti, A. Follesdal & D. Stolle (Eds.), *Politics, products, and markets* (pp. 223-244). London, UK: Transaction Publishers.

Halkier, B. (2009a). A practice theoretical perspective on everyday dealings with environmental challenges of food consumption. *Anthropology of Food, S5 September 2009.*

Halkier, B. (2009b). Suitable cooking? Performances and positionings in cooking practices among Danish women. *Food, Culture and Society, 12*, 357-377.

Halkier, B. (2010). Focus groups as social enactments: integrating interaction and content in the analysis of focus group data. *Qualitative Research, 10*(1), 71-89.

Hand, M., & Shove, E. (2004). Orchestrating Concepts: Kitchen Dynamics and Regime Change in Good Housekeeping and Ideal Home, 1922-2002. *Home Cultures, 1*(3), 1-22.

Hargreaves, T. (2008). *Making pro-environmental behaviour work: an ethnographic case study of practice, process and power in the workplace.* PhD thesis. Norwich, UK: University of East Anglia.

Hartog, A.P., den, Jobse-Putten, van, J., Otterloo, A.H., van, & Jansen, M. (1992). *De gestampte pot; Eetcultuur in Nederland.* Utrecht, the Netherlands: Nederlands Centrum voor Volkscultuur.

He, C., & Mikkelsen, E. (2009). Organic School meals in three Danish municipalities *iPOPY Discussion Paper 2/2009 Vol. 4 No. 6.*

Heerkens, N. (2009). *Stimuleren door communiceren; Stakeholder-interactie rondom duurzaam inkopen als instrument om duurzaam ondernemen binnen het MKB te stimuleren.* MSc thesis. Wageningen, the Netherlands: Wageningen University.

Hendriks, C.J.M., Stobbelaar, D.J., Fruithof, F., & Tress, B. (2004). Biologische producten met een gezicht; mogelijkheden voor regionale biologische productie om klanten te binden door herkenbaarheid. Wageningen, the Netherlands: Alterra en Leerstoelgroep Landgebruiksplanning WU.

Hermens, R. (2012). Duurzame streekproducten in de zorg (pp. 44). Oosterhout, the Netherlands: Stichting Landwaard.

Hertwich, E. (2006). *Towards a concrete sustainable consumption policy: What can we learn from examples?* Paper presented at the SCORE Perspectives on Radical Changes to Sustainable Consumption and Production (SCP), Workshop Thursday 20 and Friday 21 April 2006 Copenhagen, Denmark.

Hjelmar, U. (2011). Consumers' purchase of organic food products. A matter of convenience and reflexive practices. *Appetite, 56*(2), 336-344.

Hobson, K. (2003). Thinking habits into action: the role of knowledge and process in questioning household consumption practices. *Local Environment, 8*(1), 95-112.

Hollander, A., den (2005). Nieuwe winkelconcepten moeten klanten binden. *Ekoland, 25,* 28-31.

Holloway, L., & Kneafsey, M. (2004). Producing-consuming food: closeness, connectedness and rurality in four 'alternative' food networks. In: L. Holloway & M. Kneafsey (Eds.), *Geographies of rural cultures and societies* (pp. 262-282). London, UK: Ashgate.

Hollywood, L.E., Cuskelly, G.J., O'Brien, M., McConnon, A., Barnett, J., Raats, M.M., & Dean, M. (2013). Healthful grocery shopping. Perceptions and barriers. *Appetite, 70*(0), 119-126.

HTC (2007). Catering insights. Almere, the Netherlands: HTC Interim & Support B.V.

IEFS (1996). A pan-EU survey of consumer attitudes to food, nutrition and health. Dublin, Ireland: Institute of European Food Studies.

ING (2013). Supermarkten houden groei vast. Sectorvisie supermarkten. Amsterdam, the Netherlands: ING Economisch Bureau.

INSnet. (2007). INSnet DuurzaamheidMonitor: Deelrapport. Wageningen, the Netherlands: Wageningen University.

Jabs, J., Devine, C., Bisogni, C., Farrell, T., Jastran, M., & Wethington, E. (2007). Trying to find the quickest way: employed mothers' constructions of time for food. *Journal of Nutrition Education and Behavior, 39*(1), 18-25.

Jabs, J., & Devine, C.M. (2006). Time scarcity and food choices: An overview. *Appetite, 47*(2), 196-204.

Jackson, P., Aguila, R.P., del, Clarke, I., Hallsworth, A., Kervenoael, R., de, & Kirkup, M. (2006). Retail restructuring and consumer choice 2. Understanding consumer choice at the household level. *Environment and Planning A, 38*(1), 47-67.

Jackson, T. (2005). Motivating sustainable consumption; A review of evidence on consumer behaviour and behavioural change. Guilford, UK: Centre for Environmental Strategy, University of Surrey and SDRN, ESRC Sustainable Technologies Programme.

Jackson, T. (2007). Towards a sociology of sustainable lifestyles. *RESOLVE Working Paper 03-07.*

Jacobsen, E., & Dulsrud, A. (2007). Will consumers save the world? The framing of political consumerism. *Journal of Agricultural and Environmental Ethics, 20*(5), 469-482.

Jager, W. (2000). *Modelling consumer behavior.* PhD thesis. Groningen, the Netherlands: University of Groningen.

Jansen, J., Schuit, A.J., & Lucht, F.J., van der (2002). Tijd voor gezond gedrag; Bevordering van gezond gedrag bij specifieke groepen. Bilthoven, the Netherlands: RIVM. Available at: http://tinyurl.com/kso5d6q.

Kamphuis, C.B.M., Giskes, K., Bruijn, G.-J., de, Wendel-Vos, W., Brug, J., & Lenthe, F.J., van (2006). Environmental determinants of fruit and vegetable consumption among adults: a systematic review. *British Journal of Nutrition, 96*(4), 620-635.

Kjaernes, U., Harvey, M., & Warde, A. (2007). *Trust in food: a comparative and institutional analysis.* London, UK: Palgrave Macmillan.

Kjaernes, U., Roe, E., & Bock, B. (2007). *Societal concerns on farm animal welfare.* Paper presented at the Second welfare quality stakeholder conference, 3-4 May 2007, Berlin, Germany.

Koens, J.F. (2006). Het Digipanel over Voeding. Utrecht, the Netherlands: Milieu Centraal.

Kreijl, C.F., van, Knaap, A.G.A.C, Busch, M.C.M, Havelaar, A.H., Kramers, P.G.N., Kromhout, D., Leeuwen, F.X.R., van, Leent-Loenen, H.M.J.A., van, Ocké M.C., & Verkleij, H. (2004). Ons eten gemeten: Gezonde voeding en veilig voedsel in Nederland. Bilthoven, the Netherlands: RIVM.

Krom, M., de (2008). Understanding consumer rationalities: consumer involvement in European food safety governance of avian influenza. *Sociologia Ruralis, 49*(1), 1-19.

Krystallis, A., & Chryssohoidis, G. (2005). Consumers' willingness to pay for organic food: Factors that affect it and variation per organic product type. *British Food Journal, 107*(5), 320-343.

Lang, T. (2004). Food Industrialisation and food power: implications for food governance *Gatekeeper Series No. 114*: iied.

Lang, T., & Heasman, M. (2004). *Food wars; the global battle for mouths, minds and markets.* London, UK: Earthscan.

Lassen, A., Thorsen, A.V., Trolle, E., Elsig, M., & Ovesen, L. (2004). Successful strategies to increase the consumption of fruits and vegetables: results from the Danish – 6 a day – Work-site Canteen Model Study. *Public Health Nutrition, 7*(02), 263-270.

LEI (2006). *Landbouw economisch bericht.* The Hague, the Netherlands: Landbouw Economisch Instituut, Wageningen UR.

LNV (2005). Kiezen voor landbouw: Een visie op de toekomst van de Nederlandse agrarische sector. Den Haag, the Netherlands: LNV.

LNV (2006). *Een goed gesprek over voedselkwaliteit.* Den Haag, the Netherlands: LNV.

LNV (2008). *Houtskoolschets Europees landbouwbeleid 2020.* Den Haag, the Netherlands: LNV.

LNV. (2009). Nota duurzaam voedsel: naar een duurzame productie en consumptie van ons voedsel. Den Haag, the Netherlands: LNV.

Lockie, S., Lyons, K., Lawrence, G., & Mummery, K. (2002). Eating 'green': motivations behind organic food consumption in Australia. *Sociologia Ruralis, 42*(1).

Lury, C. (2004). *Brands, the logos of the global economy.* Cambridge, UK: Routledge.

Lusk, J.L., & Briggeman, B.C. (2009). Food values. *American Journal of Agricultural Economics, 91*(1), 184-196.

Macnaghten, P. (2003). Embodying the environment in everyday life practices1. *The Sociological Review, 51*(1), 63-84.

Makatouni, A. (2002). What motivates consumers to buy organic food in the UK? Results from a qualitative study. *British Food Journal, 104*(3/4/5), 345-352.

Marsden, T. (1998). New rural territories: regulating the differentiated rural spaces. *Journal of Rural Studies, 14*(1), 107-117.

Marsden, T., Banks, J., & Bristow, G. (2000). Food supply chain approaches: exploring their role in rural development *Sociologia Ruralis, 40*(4), 424-438.

Maubach, N., Hoek, J., & McCreanor, T. (2009). An exploration of parents' food purchasing behaviours. *Appetite, 53*(3), 297-302.

Meer, E., van der (2010). Commercieel cateren: een vak appart? Kompas voor Beleid Contractcatering. *Facto Magazine, 2010*(7/8), 18-20.

Melita, F. (2001). Organic farming in the Netherlands 2001. Frankfurt am Main, Germany: FiBL.

Memery, J., Megicks, P., Angell, R., & Williams, J. (2012). Understanding ethical grocery shoppers. *Journal of Business Research, 65*(9), 1283-1289.

Micheletti, M. (2003). *Political virtue and shopping*. New York, NY, USA: Palgrave, Macmillan.

Mikkelsen, B.E., Bruselius-Jensen, M., Andersen, J.S., & Lassen, A. (2005). Are green caterers more likely to serve healthy meals than non-green caterers? Results from a quantitative study in Danish worksite catering. *Public Health Nutrition, 9*, 1-5.

Mikkelsen, B.E., Kristensen, N.H., & Nielsen, T. (2002). Organic foods in catering – the Nordic perspective. Glostrup, Denmark: Danish Veterinary and Food Administration, Nutrition Department – The Danish Catering Centre.

Mintz, S. (1994). Eating and being: what food means. In: B. Harriss-White (Ed.), *Food: multidisciplinary perspectives* (pp. 102-115). Cambridge, UK: Basil Blackwell.

Mol, A.P.J., & Sonnenfeld, D.A. (2000). Ecological modernisation around the world; an introduction. *Environmental Politics, 9*(1), 3-14.

Morgan, D.L. (1988). *Focus groups as qualitative research* (Vol. 16). Newbury Park, CA, USA: Sage Publications.

Morgan, K. (2008). Greening the realm: sustainable food chains and the public plate. *Regional Studies, 42*(9), 1237-1250.

Morgan, K., & Morley, A. (2002). Relocalising the food chain: the role of public procurement. Cardiff, UK: The Regeneration Institute, Cardiff University.

Morgan, K., & Sonnino, R. (2007). Empowering consumers: the creative procurement of school meals in Italy and the U.K. *International Journal of Consumer Studies, 31*, 19-25.

Nationale Jeugdraad (2008). Wat is gezondheid? Onderzoeksverslag Jeugdraadpanel. Utrecht, the Netherlands: Nationale Jeugdraad.

Nelson, M., Evens, B., Bates, B., Church, S., & Boshier, T. (2007). Low income diet and nutrition; summary of main findings. London, UK: Food Standards Agency.

Oosterveer, P. (2005). *Global food governance*. PhD thesis. Wageningen, the Netherlands: Wageningen University.

Oskam, A., Meester, G., & Silvis, H. (2005). *EU policies for agriculture, food and environment*. Wageningen, the Netherlands: Wageningen Academic Publishers.

Otterloo, A.H., van (1990). *Eten en eetlust in Nederland (1840-1990)*. Amsterdam, the Netherlands: Uitgeverij Bert Bakker.

Otterloo, A.H., van (2000). Voeding in verandering. In: J.W. Schot, H.W. Lintsen, A. Rip & A.A. Bruhèze, de la (Eds.), *Techniek in Nederland in de twintigste eeuw III Landbouw voeding*. Zutphen, the Netherlands: Walburg Pers.

Otterloo, A.H., van, & Bruhèze, A.A., de la (2002). Nieuwe etenswaren, nieuwe eetplekken en veranderende eetgewoonten: snacks en fastfood in Nederland in de twintigste eeuw. In: M. Jacobs & P. Scholliers (Eds.), *Buitenshuis eten in de Lage Landen sinds 1800*. Brussels, Belgium: VUBPress.

Otterloo, A.H., van, & Sluyer, B. (2000). Naar variatie en gemak 1960-1990. In: J.W. Schot, H. Lintsen, A. Rip & A.A. Bruhèze, de la (Eds.), *Techniek in Nederland in de Twintigste Eeuw III Landbouw Voeding*. Zutphen, the Netherlands: Walburg Pers.

Otters, J. (2008). *Een overzicht van community supported agriculture in Nederland.* BSc thesis. Wageningen, the Netherlands: Wageningen University.

Owens, S. (2000). 'Engaging the public': information and deliberation in environmental policy. *Environment and Planning A, 32*(7), 1141-1148.

Padel, S., & Foster, C. (2005). Exploring the gap between attitides and behaviour: Understanding why consumers buy or do not buy organic food. *British Food Journal, 107*(8), 606-625.

Pallant, J. (2005). *SPSS survival manual* (2nd ed.). Maidenhead, UK: Open University Press.

PBL (2008). Vleesconsumptie en Klimaatbeleid. Bilthoven, the Netherlands: PBL.

Pollan, M. (2008). *In defence of food: an eater's manifesto.* New York, NY, USA: Penguin Press.

Post, A., Shanahan, H., & Jonsson, L. (2008). Food processing: barriers to, or opportunities for, organic foods in the catering sector? *British Food Journal 110*(2), 160-173

Reckwitz, A. (2002). Toward a theory of social practices: a development in culturalist theorizing. *European Journal of Social Theory, 5*(2), 243-263.

Regmi, A., Gehlhar, M., Wainio, J., Vollrath, T., Johnston, P., & Kathuria, N. (2005). Market access for high-value foods. Washington, DC, USA: USDA.

Reinders, M., Zimmermann, K., & Berg, I., van den (2009). Bedrijfsrestaurant als springplank: acceptatie van nieuwe biologische producten door introductie in de catering. Den Haag, the Netherlands: LEI Wageningen UR.

Renting, H., Marsden, T., & Banks, J. (2003). Understanding alternative food networks: exploring the role of short food supply chains in rural development. *Environment and Planning A, 35*, 393-411.

Ritzer, G. (1993). *The McDonaldisation of society.* Thousand Oaks, CA, USA: Pine Forge Press/SAGE publications.

RIVM (2010). Nationaal kompas volksgezondheid, versie 4.1, 23 september 2010. Bilthoven, the Netherlands: RIVM.

Rockeach, M.J. (1973). *The nature of human values.* New York, NY, USA: Free Press.

Roep, D., & Wiskerke, J.S.C. (2012). Reshaping the foodscape: the role of alternative food networks. In: G. Spaargaren, P.J.M. Oosterveer & A. Loeber (Eds.), *Food practices in transition: changing food consumption, retail and production in the age of reflexive modernity* (pp. 207-228). New York, NY, USA: Routledge.

Romani, S. (2005). *Feeding post-modern families: food preparation and consumption practices in new family structures.* Paper presented at the European Association for Consumer Research Conference 2005, Goteborg, Sweden, June 15-18.

Roos, E., Sarlio-Lähteenkorva, S., & Lallukka, T. (2004). Having lunch at a staff canteen is associated with recommended food habits. *Public Health Nutrition, 7*(01), 53-61.

Rutte, G., & Koning, J. (1998). *De supermarkt: 50 jaar geschiedenis.* Baarn, the Netherlands: De Prom.

Ryan, I., Cowan, C., McCarthy, M., & O'Sullivan, C. (2004). Segmenting Irish food consumers using the food-related lifestyle instrument. *Journal of International Food & Agribusiness Marketing, 16*(1), 89-114.

Schatzki, T.R. (1996). *Social practices: a Wittgensteinian approach to human activity and the social.* Cambridge, UK: Cambridge University Press.

Schatzki, T.R. (2001). *The practice turn in contemporary theory.* London, UK: Routledge.

Schulze, G. (1992). *Die Erlebnisgesellschaft; Kultursoziologie der Gegenwart.* Frankfurt, Germany: Campus Verlag.

Schwartz-Cowan, R. (1983). *More work for mother.* New York, NY, USA: Basic Books Inc.

Schwartz Cowan, R. (1987). The consumption junction: A proposal for research strategies on the sociology of technology. In: W.E. Bijker, T.P. Hughes & T.J. Pinch (Eds.), *The social construction of technological systems; New directions in the sociology and history of technology* (pp. 261-280). London, UK: The Guilford Press.

Schwartz, S.H. (1992). Universals in the content and structure of values: theoretical advances and empirical tests in 20 countries. *Advances in Experimental Social Psychology, 25*, 1-65.

SenterNovem (2010). *Startbijeenkomst actualisatie criteriadocument Catering Stakeholders Duurzaam Inkopen productgroep Catering*. Utrecht, the Netherlands: SenterNovem.

SER (2002). Innovatie voor duurzaam voedsel en groen. The Hague, the Netherlands: SER. Available at: http://tinyurl.com/odln423.

SER (2003). Duurzaamheid vraagt om openheid: op weg naar een duurzame consumptie. The Hague, the Netherlands: SER. Available at: http://tinyurl.com/qe9thk7.

Shepherd, R., Magnusson, M., & Sjödén, P.-O. (2005). Determinants of consumer behavior related to organic foods. *AMBIO, 34*(4), 352-359.

Shove, E. (2003). *Comfort, cleanliness and convenience: the social organization of normality*. Oxford, UK: Berg.

Shove, E. (2010). Beyond the ABC: climate change policy and theories of social change. *Environment and Planning A, 42*(6), 1273-1285.

Shove, E. (2012a). Habits and their creatures. In: A. Warde & D. Southerton (Eds.), *The habits of consumption* (Vol. 12, pp. 100-113). Helsinki, Finland: Collegium.

Shove, E. (2012b). Putting practice into policy: reconfiguring questions of consumption and climate change. *Contemporary Social Science 2014*, 1-15.

Shove, E., & Pantzar, M. (2005). Consumers, producers and practices. *Journal of Consumer Culture, 5*(1), 43-64.

Shove, E., & Southerton, D. (2000). Defrosting the freezer: from novelty to convenience. *Journal of Material Culture, 5*(3), 301-319.

Shove, E., & Walker, G. (2010). Governing transitions in the sustainability of everyday life. *Research Policy, 39*(4), 471-476.

Shove, E., & Warde, A. (1998). Inconspicuous consumption: the sociology of consumption and the environment. Bailrigg, UK: Department of Sociology, Lancaster University.

Sijtsema, S., Linnemann, A., Gaasbeek., T., van, Dagevos, H., & Jongen, W. (2002). Variables influencing food perception reviewed for consumer-oriented product development. *Critical Reviews in Food Science and Nutrition, 42*(6), 565-581.

Sluijter, B. (2007). *Kijken is grijpen: zelfbedieningswinkels, technische dynamiek en boodschappen doen in Nederland na 1945*. Eindhoven, the Netherlands: Technische Universiteit Eindhoven.

Sonnino, R. (2009). Quality food, public procurement, and sustainable development: the school meal revolution in Rome. *Environment and Planning A, 41*(2), 425-440.

Southerton, D., Chappells, H., & Vliet, B. van (Eds.). (2004). *Sustainable consumption: the implications of changing infrastructures of provision*. Cheltenham, UK: Edward Elgar.

Southerton, D.D. (2006). Analysing the temporal organisation of daily life: social constraints, practices and their allocation. *Sociology, 40*(3), 435.

Spaargaren, G. (1997). *The ecological modernization of production and consumption; essays in environmental sociology*. PhD thesis. Wageningen, the Netherlands: Wageningen University.

Spaargaren, G. (2011). Theories of practices: agency, technology, and culture: Exploring the relevance of practice theories for the governance of sustainable consumption practices in the new world-order. *Global Environmental Change, 21*(3), 813-822.

Spaargaren, G., & Koppen, C.S.A.K. (2009). Provider strategies and the greening of consumption practices: exploring the role of companies in sustainable consumption. In: L. Meier, & H. Lange (Eds.) *The new middle classes* (pp. 81-100). Heidelberg, Germany: Springer.

Spaargaren, G., & Mol, A.P.J. (2008). Greening global consumption: redefining politics and authority. *Global Environmental Change, 18*(3), 350-359.

Spaargaren, G., Mommaas, H., Burg, S., van den, Maas, L., Drissen, E., Dagevos, H., & Sargant, E. (2007a). More sustainable lifestyles and consumption patterns. a theoretical perspective for the analysis of transition processes within consumption domains. Wageningen, the Netherlands: Environmental Policy Group, Wageningen University.

Spaargaren, G., Mommaas, H., Burg, S., van den, Maas, L., Drissen, E., Dagevos, H., & Sargant, E. (2007b). Sustainable lifestyles and consumption patterns. Wageningen, the Netherlands: Wageningen University.

Spaargaren, G., & Oosterveer, P. (2010). Citizen-consumers as agents of change in globalizing modernity: the case of sustainable consumption. *Sustainability, 2*, 1887-1908.

Spaargaren, G., Oosterveer, P., & Loeber, A. (Eds.). (2012). *Food practices in transition: changing food consumption, retail and production in the age of reflexive modernity*. New York, NY, USA: Routledge.

Spaargaren, G., Koppen, C.S.A., van, Janssen, A.M., Hendriksen, A., & Kolfschoten, C.J. (2013). Consumer responses to the carbon labelling of food: a real life experiment in a canteen practice. *Sociologia Ruralis, 53*(4), 432-453.

Spaargaren, G., & Vliet, B., van (2000). Lifestyles, consumption and the environment: The ecological modernization of domestic consumption. *Environmental Politics, 9*(1), 50-76.

Steptoe, A., Pollard, T.M., & Wardle, J. (1995). Development of a measure of the motives underlying the selection of food: the food choice questionnaire'. *Appetite, 25*(3), 267-284.

Strasser, S. (1982). *Never done: a history of American housework*. New York, NY, USA: Pantheon Books.

Sustain (2005). Sustainable food procurement in schools, recommendations from London Food Link. Available at: http://www.sustainweb.org/pdf/Proc%20_rep_short.pdf.

Tacken, G.M.L., Winter, M.A., de, Veggel, R., van, Sijtsema, S.J., Ronteltap, A.L.C., & Reinders, M. (2010). Voorbij het broodtrommeltje; hoe jongeren denken over voedsel. Den Haag, the Netherlands: LEI.

Tanner, C., & Kast, S.W. (2003). Promoting sustainable consumption: determinants of green purchases by Swiss consumers. *Psychology and Marketing, 20*(10), 883-902.

Taylor, C. (1985[1971]). What is human agency. *Human Agency and Language, Philosophical Papers 1*, 15-44.

Thogersen, J. (2001). Consumer values, behaviour and sustainable development. *Asia Pacific Advances in Consumer Research, 4*, 207-209.

Torjusen, H., Sangstad, L., O'Dorethy Jensen, K., & Kjaernes, U. (2004). European consumers' consumptions of organic food. Oslo, Norway: National Institute for Consumer Research.

VENECA (2007). Cateringdiensten Europees aanbesteden: zo kunt u het eenvoudig zelf. Gorinchem, the Netherlands: VENECA.

Verbeek, D. (2009). *Sustainable tourism mobilities*. PhD thesis. Tilburg, the Netherlands: University of Tilburg.

Vermeir, I., & Verbeke, W. (2006). Sustainable food consumption; exploring the consumer 'attitude-behavioural intention' gap. *Journal of Agricultural and Environmental Ethics, 19*, 169-194.

Vermeir, I., & Verbeke, W. (2008). Sustainable food consumption among young adults in Belgium: theory of planned behaviour and the role of confidence and values. *Ecological Economics, 64*(3), 542-553.

VROM (2010). *Criteria voor duurzaam inkopen van catering: Versie 1.3*. Den Haag, the Netherlands: VROM.

Warde, A. (1997). *Consumption, food and taste: culinary antinomies and commodity culture*. London, UK: Sage.

Warde, A. (1999). Convenience food: space and timing. *British Food Journal, 101*(7), 518-527.

Warde, A. (2004a). Practice and field: revising Bourdieusian concepts. *CRIC Discussion Paper No 65*. Manchester, UK: The University of Manchester, Centre for Research on Innovation and Competition.

Warde, A. (2004b). Theories of practice as an approach to consumption. *Cultures of Consumption; Working Paper Series*. Manchester, UK: University of Manchester.

Warde, A. (2005). Consumption and theories of practice. *Journal of Consumer Culture, 5*(2), 131-153.

Warde, A. (2013a). Sustainable consumption and behaviour change. Available at: http://tinyurl.com/lkgv6cg.

Warde, A. (2013b). What sort of a practice is eating? In: *Sustainable practices: social theory and climate change* (pp. 17-30). Londen, UK: Routledge.

Weatherell, C., Tregear, A., & Allinson, J. (2003). In search of the concerned consumer: UK public perceptions of food, farming and buying local. *Journal of Rural Studies, 19*(2), 233-244.

Wertheim-Heck, S. (2005). Bio-logisch?! In the eye of the beholder. Den Haag, the Netherlands: LEI.

Westert, G.P., Berg, M.J., van den, Koolman, X., & Verkleij, H.E. (2008). Dutch health care performance report 2008. Bilthoven, the Netherlands: RIVM.

WHO. (2002). The world health report; reducing risks, promoting health. Geneva, Switzerland: World Health Organization.

Wilk, R. (2006). Family food fights. *Appetite, 47*(3), 401-401.

Winkel, E. (2012). *De aarde zal weer vruchtbaar zijn.* Utrecht, the Netherlands: AnkHermes.

Winter, M.A., de, Zimmermann, K.L., & Danse, M.G. (2008). Creating green consumer loyalty; How to strategically market CSR and obtain consumer preference. Den Haag, the Netherlands: LEI.

WUR. (2008). Eten van waarde. Voedselkwaliteit in Nederland. Wageningen, the Netherlands: Agrotechnology and Food Sciences Group, WUR.

Zanoli, R., & Naspetti, S. (2002). Consumer motivations in the purchase of organic food: a means-end approach. *British Food Journal, 104*(8), 643-653.

# Appendix 1. Vragenlijst voor domein voeden

In onderstaande vragen wordt gesproken over 'duurzaam voedsel'. Daarmee wordt bedoeld dat bij de productie van het voedsel het milieu en de natuur zo veel mogelijk worden ontzien en dat rekening wordt gehouden met het welzijn van dieren. Voor 'duurzaam voedsel' uit ontwikkelingslanden geldt bovendien dat het geproduceerd is onder goede arbeidsomstandigheden en dat de boeren een eerlijke prijs ontvangen voor hun producten.

## 1. Consumer concerns voeden

### *VC 1-11 Domein attitude*

In hoeverre bent u het eens met de volgende uitspraken? (Antwoordmogelijkheden: helemaal mee eens, mee eens, neutraal, mee oneens, helemaal mee oneens, weet niet/geen mening)

VC.1 De gangbare voedselproducten in winkels zijn al duurzaam genoeg
VC.2 Milieu heeft niets te maken met het voedsel dat ik in de winkel koop
VC.3 Met de technische vooruitgang zullen de gangbare voedselproducten vanzelf duurzamer worden
VC.4 De overheid en het bedrijfsleven zouden ervoor moeten zorgen dat voedsel duurzaam is
VC.5 Ik vind het belangrijk dat voedselproducten duurzaam zijn
VC.6 Ik vind het een heel gedoe om er zelf op te letten dat voedselproducten duurzaam zijn
VC.7 Ik let er op dat mijn voedselproducten duurzaam zijn
VC.8 Consumenten zijn mede verantwoordelijk voor milieuproblemen die veroorzaakt worden door de voedselproducten die ze kopen
VC.9 Consumenten zouden actief moeten kiezen voor duurzame alternatieven (bijv. biologische, Fair Trade producten, enz.)
VC.10 Voor echt duurzame voedselproducten moet je niet in de supermarkt zijn maar bij speciale winkels zoals natuurvoedingswinkels en boerenwinkels
VC.11 Om voedsel duurzaam te maken moeten landbouw en voedselproductie kleinschaliger worden

### *VC 12-18 Consumer concerns*

Welke van de volgende zaken vindt u belangrijk? (Antwoordmogelijkheden: wel belangrijk, niet belangrijk, nooit over nagedacht):

VC.12 Dat voedselproducten natuurlijk zijn (dus: onbespoten, geen kunstmatige toevoegingen en niet genetisch veranderd)
VC.13 Dat voedselproducten de natuur en het milieu niet schaden
VC.14 Dat voedselproducten niet over lange afstand zijn vervoerd
VC.15 Dat voedselproducten op een ambachtelijke manier gemaakt zijn

VC.16 Dat vlees op een diervriendelijke manier geproduceerd is
VC.17 Dat de voedselproductie bijdraagt aan de bescherming van het Nederlandse landschap en de leefbaarheid van het platteland
VC.18 Dat boeren in ontwikkelingslanden een eerlijke prijs krijgen voor hun producten en dat de arbeidsomstandigheden er goed zijn

### *VC 19-28 Oplossingsrichtingen*

Hieronder ziet u een lijst met manieren om eetgewoonten duurzamer te maken. Hoe aantrekkelijk vindt u elke manier? (Antwoordmogelijkheden: 1 = heel onaantrekkelijk, 2 = onaantrekkelijk, 3 = neutraal, 4 = aantrekkelijk, 5 = heel aantrekkelijk, 6 = weet niet/geen mening)

VC.19 Groenten en fruit van het seizoen kopen (dus bijv. geen aardbeien in de winter)
VC.20 Groenten en fruit kopen uit mijn eigen regio of uit Nederland (dus: voedsel dat niet over lange afstand is vervoerd)
VC.21 Zuinig omgaan met voedsel (dus: geen voedsel weggooien)
VC.22 Verpakking verminderen (dus bijv versproducten zonder plastic verpakking kopen, of producten met een verpakking die makkelijk in de natuur afbreekt)
VC.23 Weinig of geen vlees eten (hoogstens 3× per week)
VC.24 Vlees kopen met een speciaal keurmerk voor diervriendelijkheid of milieu (bijv. scharrelvlees, enz.)
VC.25 Biologische voedselproducten kopen (dus met het Eko keurmerk)
VC.26 Voedselproducten kopen met een milieukeurmerk
VC.27 Duurzaam gekweekte of duurzaam gevangen vis kopen (bijv. met een MSC keurmerk, of aanbevolen door Stichting Noordzee)
VC.28 Voedselproducten kopen met een keurmerk voor eerlijke handel (bijv. Fair Trade, Max Havelaar, Oké, enz.)

### *VC 29-38 De aantrekkelijkheid hiervan*

! Vervolgvragen worden gegeven voor de drie oplossingsrichtingen waarop men het hoogst scoort: U hebt aangegeven dat u het aantrekkelijk vindt om eetgewoonten duurzamer te maken door [optie] Waarom vindt u deze manier aantrekkelijk? *U mag zoveel antwoorden kiezen als u wil.*

| | |
|---|---|
| VC.29 Groenten en fruit van het seizoen kopen (bijv. geen aardbeien in de winter) | a. Is natuurlijker;<br>b. Is beter voor het milieu;<br>c. Is betrouwbaarder;<br>d. Is van betere kwaliteit;<br>e. Is lekkerder;<br>f. Is gezonder;<br>g. Is beter voor het behoud van het plattelandsleven in Nederland;<br>h. Is beter voor boeren in ontwikkelingslanden;<br>i. Anders, namelijk: |

| | |
|---|---|
| VC.30 Groenten en fruit kopen uit de eigen regio of uit Nederland (dus: voedsel dat niet over lange afstand is vervoerd) | a. Is natuurlijker;<br>b. Is beter voor het milieu;<br>c. Is betrouwbaarder;<br>d. Is van betere kwaliteit;<br>e. Is lekkerder;<br>f. Is gezonder;<br>g. Is beter voor het behoud van het plattelandsleven in Nederland;<br>h. Is beter voor boeren in ontwikkelingslanden;<br>i. Anders, namelijk: |
| VC.31 Zuinig omgaan met voedsel (geen voedsel weggooien) | a. Is natuurlijker;<br>b. Is beter voor het milieu;<br>c. Is gezonder;<br>d. Is beter voor boeren in ontwikkelingslanden;<br>e. Anders, namelijk: |
| VC.32 Verpakking verminderen (door bijvoorbeeld versproducten zonder plastic verpakking te kopen, of producten met een verpakking die makkelijk in de natuur afbreekt) | a. Is natuurlijker;<br>b. Is beter voor het milieu;<br>c. Komt de kwaliteit van het product ten goede;<br>d. Is lekkerder;<br>e. Is gezonder;<br>f. Is beter voor boeren in ontwikkelingslanden;<br>g. Anders, namelijk: |
| Weinig of geen vlees eten (hoogstens 3× per week) | a. Is natuurlijker;<br>b. Is beter voor het milieu;<br>c. Is lekkerder;<br>d. Is gezonder;<br>e. Is diervriendelijker;<br>f. Is beter voor het behoud van het plattelandsleven in g. Nederland;<br>g. Is beter voor boeren in ontwikkelingslanden;<br>h. Anders, namelijk: |
| VC.33 Vlees kopen met een speciaal keurmerk (bijv. scharrel vlees, enz.) | a. Is natuurlijker;<br>b. Is beter voor het milieu;<br>c. Is betrouwbaarder;<br>d. Is van betere kwaliteit;<br>e. Is lekkerder;<br>f. Is gezonder;<br>g. Is diervriendelijker;<br>h. Is beter voor het behoud van het plattelandsleven in Nederland;<br>i. Is beter voor boeren in ontwikkelingslanden;<br>j. Anders, namelijk: |

| | |
|---|---|
| VC.34 Biologische voedselproducten kopen (met het Eko keurmerk) | a. Is natuurlijker;<br>b. Is beter voor het milieu;<br>c. Is betrouwbaarder;<br>d. Is van betere kwaliteit;<br>e. Is lekkerder;<br>f. Is gezonder;<br>g. Is diervriendelijker;<br>h. Is beter voor het behoud van het plattelandsleven in Nederland;<br>i. Is beter voor boeren in ontwikkelingslanden;<br>j. Anders, namelijk: |
| VC.35 Voedselproducten kopen met een milieu keurmerk | a. Is natuurlijker;<br>b. Is beter voor het milieu;<br>c. Is betrouwbaarder;<br>d. Is van betere kwaliteit;<br>e. Is lekkerder;<br>f. Is gezonder;<br>h. Is beter voor het behoud van het plattelandsleven in Nederland;<br>i. Is beter voor boeren in ontwikkelingslanden;<br>j. Anders, namelijk: |
| VC.36 Duurzaam gekweekte of duurzaam gevangen vis kopen (bijv. met een MSC keurmerk, of aanbevolen door Stichting Noordzee) | a. Is natuurlijker;<br>b. Is beter voor het milieu;<br>c. Is beter voor de visstand;<br>d. Is betrouwbaarder;<br>e. Is van betere kwaliteit;<br>f. Is lekkerder;<br>g. Is gezonder;<br>h. Is beter voor het behoud van het plattelandsleven in Nederland;<br>i. Is beter voor boeren in ontwikkelingslanden;<br>j. Anders, namelijk: |
| VC.37 Voedselproducten kopen met een keurmerk voor eerlijke handel (bijv. Fair Trade, Max Havelaar, Oké, enz.) | a. Is natuurlijker;<br>b. Is beter voor het milieu;<br>c. Is betrouwbaarder;<br>d. Is van betere kwaliteit;<br>e. Is lekkerder;<br>f. Is gezonder;<br>g. Is diervriendelijker;<br>h. Is beter voor het behoud van het plattelandsleven in Nederland;<br>i. Is beter voor boeren in ontwikkelingslanden;<br>j. Anders, namelijk: |

## 2. Portfolio's voeden

### *VP 1-8 Betreft een typering van de gedragspraktijk voor duurzame voedsel consumptie aan de hand van via welke kanalen de respondent aan duurzame voedselproducten komt*

De volgende vragen gaan over uw 'boodschappengedrag'.
In hoeverre kloppen de volgende uitspraken voor uw huishouden? (Antwoordmogelijkheden: helemaal op mij van toepassing, beetje op mij van toepassing, neutraal, niet erg op mij van toepassing, helemaal niet op mij van toepassing, weet niet/geen mening)

Bij mij/bij ons worden:

VP.1 De boodschappen meestal bij dezelfde supermarkt gedaan
VP.2 De boodschappen bij verschillende supermarkten gedaan
VP.3 De boodschappen bij andere winkels dan de supermarkt gedaan
VP.4 De boodschappen gedaan bij de supermarkt of winkel waar ze aantrekkelijke aanbiedingen hebben
VP.5 De boodschappen gedaan bij de groentewinkel/groenteboer, de slager of de bakker, (dus: bij speciaalzaken)
VP.6 De boodschappen gedaan bij de natuurvoedingswinkel of biologische supermarkt
VP.7 Een deel van de boodschappen gedaan bij de boer (in een boerderijwinkel of op een boerenmarkt)
VP.8 De boodschappen gedaan via het internet

### *VP 9-14 Kennis domein voeden*

Hoeveel weet u over de volgende onderwerpen?
(Likertschaal: weet ik niets van, weet ik weinig van, weet ik wel iets van, weet ik veel van)

VP.9 Hoe je maaltijden duurzamer kunt maken door met andere ingrediënten te koken
VP.10 De milieuproblemen die door de landbouw worden veroorzaakt
VP.11 De milieuproblemen die door de vlees- en voedingsindustrie worden veroorzaakt
VP.12 Het uitsterven van bepaalde vissoorten door de grootschalige visvangst
VP.13 Welke voedselproducten duurzaam zijn en welke niet
VP.14 De problemen in de derde wereld die veroorzaakt worden door de handel in voedsel

## *VP 15-23 Eigen handelingservaring duurzaam consumeren voeden*

De eerder genoemde manieren om eetgewoonten duurzamer te maken komen nu weer aan bod. Wilt u aangeven welke dingen u zelf doet, hoe vaak, en in welke situatie u er op let.

| | Optie | Hoe vaak | In welke situatie(s) |
|---|---|---|---|
| VP.15 | Ik koop groenten en fruit van het seizoen | Bijna altijd – regelmatig – af en toe – nooit, maar ik weet wel wanneer welke producten in het seizoen zijn – nooit en ik weet ook niet wanneer welke producten in het seizoen zijn | Ik let er op dat groenten en fruit van het seizoen zijn:<br>a. tijdens het boodschappen doen;<br>b. in de kantine op mijn werk;<br>c. bij het eten in horeca gelegenheden;<br>d. geen van deze. |
| VP.16 | Ik koop groenten en fruit kopen uit mijn eigen regio of uit Nederland (dus: voedsel dat niet over lange afstand is vervoerd) | Bijna altijd – regelmatig – af en toe – nooit, maar ik weet wel wanneer welke producten in het seizoen zijn – nooit en ik weet ook niet wanneer welke producten in het seizoen zijn | Ik let er op dat groenten en fruit uit mijn eigen regio of uit Nederland komen:<br>a. tijdens het boodschappen doen;<br>b. in de kantine op mijn werk;<br>c. bij het eten in horeca gelegenheden;<br>d. geen van deze. |
| VP.17 | Ik ga zuinig om met voedsel (probeer geen voedsel weg te gooien) | Bijna altijd – regelmatig – af en toe – nooit | Ik let er op dat ik zuinig omga met voedsel:<br>a. thuis;<br>b. in de kantine op mijn werk;<br>c. bij het eten in horeca gelegenheden;<br>d. geen van deze. |
| | Ik koop versproducten waar zo weinig mogelijk verpakking omheen zit; | Bijna altijd – regelmatig – af en toe – nooit | Ik let op de verpakking:<br>a. tijdens het boodschappen doen;<br>b. in de kantine op mijn werk;<br>c. bij het eten in horeca gelegenheden;<br>d. geen van deze. |
| VP.18 | Ik koop vaker producten waar een composteerbare verpakking omheen zit;<br>Keurmerk voor composteerbare verpakking (die afbreekt in de natuur) (afbeelding keurmerk) | Bijna altijd – regelmatig – af en toe – nooit maar ik ken die verpakkingen wel, nooit en ik ken die verpakkingen ook niet | |
| VP.19 | Ik eet weinig of geen vlees (hoogstens 3× per week); | Bijna altijd – regelmatig – af en toe – nooit | |

| | | | |
|---|---|---|---|
| VP.20 | Ik koop vlees met een speciaal keurmerk voor diervriendelijkheid en/of milieu.<br>**Toelichting:**<br>Bij vlees zien we een aantal verschillende keurmerken die aangeven dat er is gelet op dierenwelzijn en milileu. Hieronder staan enkele voorbeelden van deze keurmerken:<br>(afbeeldingen keurmerken) | Bijna altijd – regelmatig – af en toe – nooit maar ik ken het wel – nooit en ik ken het ook niet | Ik let op vlees keurmerken:<br>a. tijdens het boodschappen doen;<br>b. in de kantine op mijn werk;<br>c. bij het eten in horeca gelegenheden;<br>d. geen van deze. |
| VP.21 | Ik koop biologische voedselproducten (met het Eko keurmerk);<br>(afbeelding keurmerk) | Bijna altijd – regelmatig – af en toe – nooit maar ik ken ze wel, nooit en ken ze ook niet | Ik let op biologisch producten:<br>a. tijdens het boodschappen doen;<br>b. in de kantine op mijn werk;<br>c. bij het eten in horeca gelegenheden;<br>d. geen van deze. |
| VP.22 | Ik koop duurzaam gekweekte vis (met het MSC keurmerk of de viscodering)<br><br>Voorbeeld:<br>MSC keurmerk (afbeelding keurmerk)<br>vis kleur-codering Stichting De Noordzee (afbeeldingen keurmerken) in de viswijzer (afbeelding viswijzer) of op het schap van de visboer | Bijna altijd – regelmatig – af en toe – nooit maar ik ken dit wel – nooit en ik ken dit ook niet | Ik let er op dat de vis duurzaam gekweekt is:<br>a. tijdens het boodschappen doen;<br>b. in de kantine op mijn werk;<br>c. bij het eten in horeca gelegenheden;<br>d. geen van deze. |
| VP.23 | Ik koop producten met een keurmerk voor eerlijke handel (zoals, Fair Trade, Max Havelaar, Utez Kapeh, e.d.)<br>(afbeelding keurmerken) | Bijna altijd – regelmatig – af en toe – nooit maar ik ken ze wel – nooit en ik ken ze niet | Ik let op 'eerlijke',producten:<br>a. tijdens het boodschappen doen;<br>b. in de kantine op mijn werk;<br>c. bij het eten in horeca gelegenheden;<br>d. geen van deze. |

## *VP 24-32 Informatie bronnen*

(Likertschaal 4: 1 = vaak, 2 = regelmatig, 3 = af en toe, 4 = nooit):
Ik krijg weleens informatie over de duurzaamheid van voedselproducten van:

VP.24 Organisaties die zich op voeding, milieu of natuurbescherming richten (bijv. Voedingscentrum, Natuur en Milieu, Greenpeace);
VP.25 Een consumentenorganisatie (bijv. Consumentenbond, Goede Waar & Co);
VP.26 Televisieprogramma's over lifestyle en/of koken;
VP.27 Het etiket op de verpakking van voedselproducten;
VP.28 Medewerkers van de winkel;

VP.29 Informatiemateriaal dat in de winkel aanwezig is;
VP.30 Tijdschriften over eten en/of koken;
VP.31 Familie, vrienden of kennissen;
VP.32 Ik ga volledig af op mijn eigen kennis en ervaring

## 3. Groene aanbodstrategieën voeden

### *VA 1-88 Evaluatie aanbod en aanbodstrategieen*

Respondenten krijgen maximaal 2 opties voorgelegd van de opties waar ze eerder 4 of 5 op scoorden (aantrekkelijkheid in VC vragen) (of waar ze in ieder geval niet 1 of 2 op scoorden).
In hoeverre bent u het eens met de volgende uitspraken?
(Antwoordmogelijkheden: 1 = helemaal mee oneens; 2 = mee oneens; 3 = neutraal; 4 = mee eens; 5 = helemaal mee eens, weet niet/geen mening)

*Optie 1: 'Groenten en fruit van het seizoen kopen (bijv. geen aardbeien in de winter)'*

U hebt aangegeven dat u dit een aantrekkelijke manier vindt om eetgewoonten duurzamer te maken.
VA.1 Ik kan gemakkelijk aan seizoensproducten komen

Als beter bekend was welke producten van het seizoen zijn,
VA.2 zou ik meer gebruik maken van seizoensproducten
VA.3 zouden de mensen meer gebruik maken van seizoensproducten

VA.4 Er zijn genoeg verschillende seizoensproducten te krijgen
VA.5 Groenten en fruit smaken beter als ze van het seizoen zijn
VA.6 Het gebruik van seizoensproducten past bij mijn stijl van koken en eten

Als beter bekend was hoe je lekker kunt koken met seizoensproducten,
VA.7 zou ik ze vaker gebruiken
VA.8 zouden de mensen vaker seizoensproducten gebruiken

VA.9 Het kopen van seizoensproducten is een goede manier om iets te doen aan duurzame voeding.

Als het duidelijker was dat seizoensproducten beter zijn voor het milieu,
VA.10 zou ik ze eerder kopen
VA.11 zouden de mensen ze vaker kopen

*Optie 2: 'Groenten en fruit kopen uit de eigen regio of uit Nederland (dus: voedsel dat niet over lange afstand is vervoerd)*

U hebt aangegeven dat u dit een aantrekkelijke manier vindt om eetgewoonten duurzamer te maken.
In hoeverre bent u het eens met de volgende uitspraken?
(Antwoordmogelijkheden: 1 = helemaal mee oneens; 2 = mee oneens; 3 = neutraal; 4 = mee eens; 5 = helemaal mee eens, weet niet/geen mening)

VA.12 Ik kan gemakkelijk aan groenten en fruit uit mijn eigen regio of uit Nederland komen
VA.13 Er zijn genoeg verschillende groenten en fruitsoorten uit mijn eigen regio of uit Nederland te krijgen
VA.14 De kwaliteit van groenten en fruit uit mijn eigen regio of uit Nederland is goed
VA.15 Ik heb meer vertrouwen in groenten en fruit uit mijn eigen regio of uit Nederland dan in groenten en fruit uit het buitenland
VA.16 Het kopen van groenten en fruit dat minder ver heeft gereisd is een goede manier om iets te doen aan duurzame voedingbeter voor het milieu

Als duidelijker stond aangegeven over hoeveel kilometers groenten en fruit zijn vervoerd,
VA.17 zou ik er op letten en groenten en fruit van dichtbij kopen
VA.18 zouden de mensen meer groenten en fruit van dichtbij kopen

*Optie 3: 'Zuinig omgaan met voedsel (geen voedsel weggooien)'*

U hebt aangegeven dat u dit een aantrekkelijke manier vindt om eetgewoonten duurzamer te maken.
In hoeverre bent u het eens met de volgende uitspraken?
(Antwoordmogelijkheden: 1 = helemaal mee oneens; 2 = mee oneens; 3 = neutraal; 4 = mee eens; 5 = helemaal mee eens, weet niet/geen mening)

VA.19 Ik kan gemakkelijk zuinig zijn met voedsel, bijv. door kleinere porties verpakte groenten te kopen, goed afsluitbare verpakkingen te gebruiken, zelf groenten in te vriezen, overgebleven eten te bewaren
VA.20 Ik vind het makkelijk om zuinig om te gaan met voedselproducten
VA.21 Ik ben het gewend zuinig om te gaan met voedselproducten
VA.22 Als ik wist hoe ik zuinig moet omgaan met voedsel zou ik er eerder op letten
VA.23 Als mensen wisten hoe ze zuinig om moeten gaan met voedsel, zouden ze er eerder op letten
VA.24 Ik vind dat mensen eigenlijk geen voedsel weg zouden moeten gooien
VA.25 Zuinig omgaan met voedsel is een goede manier om iets te doen aan duurzame voeding

*Optie 4: 'Versproducten kopen zonder plastic verpakking of met goed afbreekbare verpakking'*

U hebt aangegeven dat u dit een aantrekkelijke manier vindt om eetgewoonten duurzamer te maken.
In hoeverre bent u het eens met de volgende uitspraken?
(Antwoordmogelijkheden: 1 = helemaal mee oneens; 2 = mee oneens; 3 = neutraal; 4 =mee eens; 5 = helemaal mee eens, weet niet/geen mening)

VA.26 Ik kan gemakkelijk aan groenten en fruit komen waar geen plastic verpakking omheen zit.
VA.27 Ik kan gemakkelijk aan producten komen waar een natuurlijk afbreekbare verpakking omheen zit.
VA.28 Ik vind de plastic verpakking om groenten en fruit niet nodig.
VA.29 Ik vind de plastic verpakking om groenten en fruit handig.
VA.30 Als de versproducten in de supermarkt/winkel minder in plastic verpakt waren, zou ik dat fijn vinden.
VA.31 Er zijn genoeg verschillende versproducten met natuurlijk afbreekbare verpakking te krijgen.

Als duidelijker was welke verpakkingen natuurlijk afbreekbaar zijn en welke niet,
VA.32 zou ik eerder versproducten met zulke verpakkingen kopen
VA.33 zouden mensen eerder versproducten met zulke verpakkingen kopen

VA.34 Letten op de verpakking is een goede manier om iets te doen aan duurzame voeding
VA.35 Ik vertrouw er op dat natuurlijk afbreekbare verpakking beter is voor het milieu dan verpakking die niet gemakkelijk afbreekt

*Optie 5: Weinig of geen vlees eten (hoogstens 3× per week)*

U hebt aangegeven dat u dit een aantrekkelijke manier vindt om eetgewoonten duurzamer te maken.
In hoeverre bent u het eens met de volgende uitspraken?
(Antwoordmogelijkheden: 1 = helemaal mee oneens; 2 = mee oneens; 3 = neutraal; 4 = mee eens; 5 = helemaal mee eens, weet niet/geen mening)

VA.36 Ik kan gemakkelijk aan vleesvervangers komen
VA.37 Ik vind vleesvervangers lekker
VA.38 Vegetarisch koken past bij mijn eigen stijl van eten en koken
VA.39 Ik vind vegetarisch eten lekker
VA.40 Als mensen beter wisten hoe ze lekkere vegetarische maaltijden kunnen klaarmaken, zouden ze dat vaker doen
VA.41 Minder vlees eten is een goede manier om iets te doen aan duurzame voeding

*Optie 6: 'Vlees kopen met een speciaal keurmerk voor dier- en milieuvriendelijk'*

U hebt aangegeven dat u dit een aantrekkelijke manier vindt om eetgewoonten duurzamer te maken.
In hoeverre bent u het eens met de volgende uitspraken?
(Antwoordmogelijkheden: 1 = helemaal mee oneens; 2 = mee oneens; 3 = neutraal; 4 = mee eens; 5 = helemaal mee eens, weet niet/geen mening)

VA.42 Ik kan gemakkelijk aan vlees komen met een keurmerk voor dier- en milieuvriendelijk vlees

Als de slager of het personeel in de supermarkt aan de klanten zouden vertellen welk vlees dier- en milieuvriendelijk is en welk vlees niet, dan
VA.43 zou ik eerder dier- en milieuvriendelijk vlees kopen
VA.44 zouden de mensen eerder dier- en milieuvriendelijk vlees kopen

Als er in de winkel een duidelijk overzicht was van de verschillende keurmerken voor vlees en hun betekenis (zie voorbeeld afbeelding met keurmerken en uitleg hierover),
VA.45 zou ik eerder vlees kopen met deze keurmerken
VA.46 zouden de mensen eerder vlees kopen met deze keurmerken

VA.47 Er zijn genoeg verschillende soorten dier- en milieuvriendelijk vlees (dus met een keurmerk van diervriendelijkheid) te krijgen
VA.48 Vlees kopen met een speciaal keurmerk over diervriendelijkheid en milieu is een goede manier om iets te doen aan duurzame voeding
VA.49 Over het algemeen vertrouw ik de keurmerken die aangeven of vlees dier- en milieuvriendelijk geproduceerd is
VA.50 Als het om dier- en milieuvriendelijk vlees gaat, heb ik vertrouwen ik de keurmerken van bedrijven (bijv. de supermarkt, de vleeshandel/industrie)
VA.51 Als het om dier- en milieuvriendelijk vlees gaat, heb ik vertrouwen in het Eko keurmerk (biologisch)
VA.52 Als het om dier- en milieuvriendelijk vlees gaat, heb ik vertrouwen in het advies van mijn slager
VA.53 Als het om dier- en milieuvriendelijk vlees gaat, heb ik vertrouwen in een keurmerk van de overheid
VA.54 Als het om dier- en milieuvriendelijk vlees gaat, heb ik meer vertrouwen in een keurmerk van de overheid dan in andere keurmerken

*Optie 7: 'Biologische voedselproducten kopen (met het Eko keurmerk)'*

U hebt aangegeven dat u dit een aantrekkelijke manier vindt om eetgewoonten duurzamer te maken.
In hoeverre bent u het eens met de volgende uitspraken?
(Antwoordmogelijkheden: 1 = helemaal mee oneens; 2 = mee oneens; 3 = neutraal; 4 = mee eens; 5 = helemaal mee eens, weet niet/geen mening)

VA.55 Ik kan gemakkelijk aan biologische producten komen.
VA.56 Biologische producten zouden standaard in alle supermarkten en winkels aanwezig moeten zijn.

Als in de winkel duidelijker werd aangegeven welke producten biologisch zijn en welke niet,
VA.57 zou ik eerder biologische producten kopen
VA.58 zouden meer mensen biologische producten kopen

Als in de winkel meer informatie (bijv. folders) aanwezig was over wat biologisch inhoudt,
VA.59 zou ik eerder biologische producten kopen
VA.60 zouden meer mensen biologische producten kopen

VA.61 Het assortiment aan biologische producten in de winkel/supermarkt is meestal goed.
VA.62 Biologische producten zijn over het algemeen van even goede kwaliteit als niet-biologische producten.
VA.63 Biologische producten passen bij mijn eigen stijl van eten en koken.

Als er een groter assortiment aan biologische producten beschikbaar was,
VA.64 ik eerder biologische producten kopen
VA.65 zouden meer mensen biologische producten kopen

VA.66 Het kopen van biologische producten is een goede manier om iets te doen aan duurzame voeding.
VA.67 Ik vertrouw erop dat producten met het Eko keurmerk duurzamer zijn dan niet-biologische producten.

Als duidelijker was dat biologische producten beter zijn voor het milieu dan niet-biologische producten,
VA.68 zou ik eerder biologische producten kopen
VA.69 zouden de mensen eerder biologische producten kopen

*Optie 8: 'Bij het kopen van vis let ik er op dat het duurzaam gevangen vis is'*

U hebt aangegeven dat u dit een aantrekkelijke manier vindt om eetgewoonten duurzamer te maken.
In hoeverre bent u het eens met de volgende uitspraken?
(Antwoordmogelijkheden: 1 = helemaal mee oneens; 2 = mee oneens; 3 = neutraal; 4 = mee eens; 5 = helemaal mee eens, weet niet/geen mening)

VA.70 Ik kan gemakkelijk aan duurzame vis komen
VA.71 Als de visboer zou adviseren welke vis duurzaam is en welke niet zouden meer mensen duurzame vis kopen
VA.72 Er zijn genoeg verschillende soorten duurzame vis te krijgen
VA.73 Ik vind de kwaliteit van vis met een duurzaamheidskeurmerk van MSC even goed als de kwaliteit van vis zonder zo'n keurmerk
VA.74 Als bij de visboer en in de schappen van de supermarkt duidelijker werd aangeven welke vis duurzaam is en welke niet zouden mensen eerder duurzame vis kopen
VA.75 Het kopen van duurzaam gevangen vis is een goede manier om iets te doen aan duurzame voeding

Als het om duurzame vis gaat,
VA.76 heb ik vertrouwen in de keurmerken van bedrijven (bijv. de supermarkt, of de vishandel)
VA.77 heb ik vertrouwen in het MSC keurmerk
VA.78 heb ik vertrouwen in de informatie van Stichting Noordzee (van de Viswijzer)
VA.79 heb ik vertrouwen in het advies van mijn visboer.
VA.80 heb ik vertrouwen in een keurmerk van de overheid

*Optie 9: 'Voedselproducten kopen met een keurmerk voor eerlijke handel (bijv. Fair Trade, Max Havelaar, Oké, enz.)'*

U hebt aangegeven dat u dit een aantrekkelijke manier vindt om eetgewoonten duurzamer te maken.
In hoeverre bent u het eens met de volgende uitspraken?
(Antwoordmogelijkheden: 1 = helemaal mee oneens; 2 = mee oneens; 3 = neutraal; 4 = mee eens; 5 = helemaal mee eens, weet niet/geen mening)

VA.81 Ik kan gemakkelijk aan 'eerlijke' voedselproducten komen
VA.82 Er zijn genoeg verschillende 'eerlijke' voedselproducten te krijgen
VA.83 Deze 'eerlijke' voedselproducten zijn van even goede kwaliteit als producten die niet zo'n keurmerk hebben
VA.84 Het kopen van 'eerlijke' producten is een goede manier om iets te doen aan duurzame voeding

VA.85 Als het om eerlijke producten gaat, heb ik vertrouwen in keurmerken zoals Fair Trade, Max Havelaar en Oké

Als in de winkels duidelijker werd aangegeven welke producten 'eerlijk' zijn en welke niet,
VA.86 zou ik deze 'eerlijke' producten eerder kopen
VA.87 zouden de mensen deze 'eerlijke' producten eerder kopen

Als er meer achtergrondinformatie in de winkel aanwezig was over hoe deze 'eerlijke' producten bijdragen aan betere werkomstandigheden voor boeren en landarbeiders,
VA.88 zou ik deze 'eerlijke' producten eerder kopen
VA.89 zouden de mensen deze 'eerlijke' producten eerder kopen

# Appendix 2. Algemene vragenlijst cateraars

## Visie en strategie

- Wat verstaat u onder 'duurzame catering'/'duurzaam voedsel' binnen uw bedrijf? (Hoe ziet u het onderwerp duurzaam in relatie tot brede aspecten van kwaliteit binnen de bedrijfscatering?)
- Hoe zou u uw aanbodstrategie(en) beschrijven t.a.v. duurzame voeding/duurzaamheid? (Zijn deze toegespitst op bepaalde consumenten/klanten(opdrachtgevers)/kantine locaties?)
- Wat zijn de belangrijkste kansen en belemmeringen dit u tegenkomt in het aanbieden van een duurzaam productaanbod?
- Hoeveel ruimte is er om duurzaam te opereren? (contract catering)
- (Hoe ziet u de rol van food professionals (cateraars/koks) in het verduurzamen van voedselpraktijken? (bijv. thema opleiding koks))

## Keten

- Van wat voor soort leveranciers maakt u gebruik (inkoop duurzame producten)?
- Hoe vindt u het aanbod van duurzame producten bij leveranciers?
- Hoe is de samenwerking tussen cateraars en leveranciers op het gebied van duurzame productaanbod?
- Is er meer vraag naar 'duurzaam' voedsel onder gebruikers/opdrachtgevers dan vier tot vijf jaar geleden? (Ziet u een stijgende lijn?)

## In de kantine

- Hoe ziet de kantine er uit?
- Kunt u iets zeggen over wat de succesformules (t.a.v. productaanbod en communicatie) zijn om een duurzamere lunch aan de man te brengen?
- Vindt u dat er genoeg wordt ingespeeld op de eindgebruiker (ook de latente behoefte)? Hoe gaat u hiermee om? (Hoe vindt u dat de terugkoppeling tussen kantine/cateraar en gebruikers moet zijn?)

# List of interviewees

| | |
|---|---|
| Alber Kooy | Stenden Catering (leerbedrijf 'Canteen') |
| Anne van den Oord | Coördinator van Eerlijk & Heerlijk/Real Food (Sligro) |
| Anton Damen | Facility management adviseur, Sense FM |
| Bob Hutten | Hutten Catering |
| Doris Vos | Tijd voor Eten |
| Edgar Zonneveldt | Dienst Milieu en Bouwtoezicht, Gemeente Amsterdam |
| Els Smit | Commissie/werkgroep duurzaamheid Veneca |
| Fer de Bruin | Werkt voor Ouderen, Ter Reede, Zeeland |
| Han Soethoudt | onderzoeker |
| Hanna Schösler | Avenance/Elior |
| Henk Voormolen, | Directeur Marketing Services, Albron |
| Ieke Benschop | Projectleider duurzame catering, Natuur en Milieufederatie Utrecht |
| Ieneke Snijders | Albron |
| J. De Jong | Stiching Milieukeur |
| Lars Charas | Voormalig kok RIVM, Bilthoven |
| Maria Voors | Maria's Biologische Eetwinkel, Utrecht |
| Nico Heukels | Project Restaurant van de Toekomst, Wageningen |
| René Hagels | Beleidsmedewerker van de Universitaire Restauratieve Dienst, Universiteit Nijmegen |
| René Koster | Project Restaurant van de Toekomst, Wageningen |
| Rijkent Cornelius | Project Biotbites |
| Ronald van den Broek | Inkoper, La Place |
| Sandra Klarenbeek | Facilitair Bedrijf, Restauratieve Voorziening Erasmus Universiteit |
| Suzanne Stienen | Specialist Buying & Merchandising DeliXL |
| Willen ter Voert | Hoofd afdeling catering, Bedrijfsrestaurant Provincie Gelderland |

# Summary

Consumers and their (potential) actions are given a central role within the realisation of (further) sustainable development of agro-food systems. On the subject of making everyday food consumption in society more sustainable, i.e. more ecologically and socially sound, much research is geared toward a better understanding of consumers' individual preferences, values and attitudes. However, this is often done with disregard to the contexts and settings consumers find themselves in and the (infra)structures of food provisioning. Studies have indicated that bad food infrastructure or 'obsogenic' environments promote unhealthy eating patterns in society (Dagevos & Munnichs, 2007; Gezondheidsraad, 2003), and that the availability and accessibility of healthy foods is paramount in promoting healthier eating patterns (Brug & Lenthe, 2005; Kamphuis *et al.*, 2006). The same is likely to be applicable to (un)sustainable eating habits. Contexts of food consumption bring forth certain consumption patterns and are 'sites' of meaning creation related to food. Taking context into account may enable us to see possibilities and opportunities which we might otherwise not.

This study focuses on food consumption as a practice, thus incorporating the situational and infrastructural dimension of consumption. Practice perspectives hold that consumption patterns or behaviours, consumer attitudes and opinions on food are shaped by and within these everyday food practices. They look to investigate the contexts in which consumption takes place, meaning the social and physical infrastructures of consumption, and the actors which have a (potential) influence on determining consumption practices. A practice approach sheds a different light on the position of the consumer within the sustainable development of the food domain and the ways in which everyday food consumption might improve in terms of sustainability.

For the purpose of this research, we take the sustainable development of food consumption to mean increasing the use of and engagement in various sustainable food alternatives, but also the improvement in provisioning of sustainable alternatives in a boarder range of consumption contexts within society, like within canteen facilities within offices and educational facilities. Firstly, this study considers the position of the consumer with the sustainable development of the agro-food sector. Secondly, it studies the extent to which Dutch consumers engage in sustainable food consumption and how consumers with different food shopping practices differ in terms of this engagement. Lastly, it considers sustainable food provisioning and consumption within a specific context outside the home; public and private canteen catering.

Today, food is eaten and appropriated on a daily basis in various different places and contexts, from the caterer at work, the sandwich bar at the train station or the petrol station, etc. Food consumption thus involves a whole pallet of different home and out-of-home food consumption practices, each with their own dynamics of food provisioning, modes of access and use, and social interactions. Especially in (sub)urban areas, the a major characteristic of food consumption is differentiation and 'flexibility' in terms of access to various food services and the use of convenient 'food solutions' to fit busy lifestyles, tight budgets, etc. and/or possibly a lack of cooking time and skills. Thus, although our food habits have become more individualistic, fitting our individual ideas, concerns, preferences and schedules, those with a big hand in shaping food values, preferences and tastes are likely to be food providers in the catering and food retail industry. The limitation to direct consumer influence and responsibility relating to 'the daily meal' is in contrast to the development seen in the general awareness and concern consumers have for food sustainability and health issues, animal welfare and rural welfare issues at home and abroad. Furthermore, the fact that the use of convenience foods and food services are so firmly embedded in everyday food practice suggests that is important that sustainable food alternatives are sufficiently available and accessible in varying settings and locations where food is ordered, shopped for and/or eaten.

The survey study aimed to explore in more detail the perception consumers have of their own responsibility (co-responsibility) within sustainable development of food consumption, and their concerns and considerations in as to actively participating in more sustainable forms of consumption. Different facets of the consumers' role are discussed; a distinction is made between the role consumers see for themselves, for consumers in general, 'in theory' and 'in practice'. Dutch consumers were questioned on their engagement in a wide range of different sustainable food alternatives applied in daily food practice (from combatting food wastage to eating less meat and buying organic food) and how this engagement differed between consumers with different everyday shopping practices.

The most important results from the survey study are that respondents who shop in alternative shops do not score much higher on co-responsibility (total score) than those who shop at supermarkets and respondents with different shopping practices do not score differently on their opinions of the link between the environment and food consumption. The differences between respondents with different shopping practices becomes more marked when the subject is applying sustainability in practice; those who shop exclusively in supermarkets are less inclined to make sure the food they buy is sustainable compared with respondents with other kinds of shopping practices. Thus, supermarket shoppers appear to be not less responsible overall, but are less likely to come home with a sustainable alternative. The most pronounced differences between shopping practices are found in respondents' use of certain product alternatives: people who shop exclusively in the supermarket make less frequent use of regional/local, sustainably produced fish, organic, eco-labelled and Fair Trade food, and meat with a label on animal/environmental care. Except for buying seasonal fruits and vegetables; this is practiced equally by those visiting supermarkets and alternatives shops. All respondents appear to engage similarly often in cutting down on food wastage and packaging. Overall, the findings illustrate that consumers with different shopping practices have different portfolios, that is, different knowledge about food sustainability and/or experience(s) with sustainable food alternatives. The differences are found especially in terms of use of certain product alternatives. Portfolios are much less different in terms of the use of practical alternatives like cutting down on food wastage and packaging and the degree to which alternatives are positively evaluated.

The study on catering revealed the bottlenecks and opportunities for further sustainable development in both food provisioning an consumption. Firstly, it illustrated the importance of contracts in determining what kind of food will be served in canteens. Resistance to sustainable development in canteen food provisioning can be caused by the inflexibility which arises from being tied to a prior contracts. It is vital that sustainability criteria are built into the individual contractual agreements with contract-lenders from the start. The caterer and the contract-lender must make clear agreements on how sustainability is to be realised. 'Creative procurement' can go a long way to encapsulate sustainability requirements, offering the opportunity to connect procurement to shared food values and/or other sustainable development goals within an organisation. Secondly, it showed that kitchen infrastructure has an important role to play in determining what kinds of options are available to those who want to 'green the canteen'. Canteens with skilled food professionals and proper 'cooking' kitchens will be better equipped to make changes in recipes and menus themselves. This allows for a degree of creativity and flexibility in realising sustainable provisioning. In cases where kitchens are absent caterers are more dependent on the availability and quality of sustainable food alternatives from suppliers and wholesalers. Thirdly, caterers often employ near-sourcing and seasonal scouring in their business strategies. Near-sourcing in many cases is not a goal in itself, but rather plays a role in enabling sustainable provisioning cost-wise, and in adding value to catering services. Many food professionals (cooks, caterers) engage in near-sourcing due to certain quality advantages like tractability and transparency, freshness and taste, or making the use of organic or other sustainably produced foods affordable. Others want to support their local economy, and see socio-economic benefits in local/regional sourcing. The rationale of food service providers lies within a practical reality; near sourcing can offer food professionals practical solutions for improving their services and realising more sustainable provisioning. Fourthly, the catering study showed that realising sustainable food procurement requires good canteen management. Competence, having the right knowledge and skill is a reoccurring theme within this chapter. It seems that realising sustainable provisioning has as much to do with the availability of sustainable food alternatives on the market as with the competence of those in charge of organisation catering services. Here we discover the importance of the role of food professionals and their capabilities. Whereas organisations (the caterers' employer in the case of in-house catering and the contract-lending party in the case of hired catering) need competence to manage catering procurement processes, capable food professionals play a key role in providing the actual service to the end-user.

The second part of the catering study revealed how food choices/practices in the canteen setting are made in relation to work and study routine. It illustrated how a person's reasoning about their food practices entail descriptions of how they manage and experience food intake in their everyday lives. Generally speaking, the employees and students who participated in the focus group expressed their support for a sustainably operating canteen and one which serves enough sustainably produced foods. However, the canteen context was considered different from their 'home-context'. Some expressed that they felt they have little influence on what is served in the canteen. Others expressed how issues of time and scheduling, tastes and fancies or other considerations play a role in their daily canteen food choice. Furthermore, judgements about the quality of the canteen food services were often related in terms of perceptions about value for money and/or past experiences with canteen food. Facility experts have expressed the need for catering services within work-place and higher-education to be more focussed on the end-user. In some cases this might be realised by involving end-users more structurally in decision making on food and canteen procurement. Other options lie in gaining a better understanding of end-users practices, concerns and wishes, and/or framing food procurement in terms of certain organisations values and/or goals. Caterers often focus on combining sustainability issues with the art of good food and good catering hospitality. In that sense they approach sustainability using the 'food experience strategy'; they combine sustainability with goals of maintaining a healthy and successful business (i.e. people, planet and profit). Here, the focus is less on information provisioning and/or creating dialogue or debate on sustainability with consumers, although this may play a role, and more on developing the food service itself. The 'food experience' strategy seems to contain the promise of being able to change every day canteen food consumption practices to some degree. Food habits and routines can be challenged through changing the canteen setting and the kind of food served there.

The concluding chapter argues that a practice based approach offers insights which go beyond the realm of abstract, individual and/or structural conditions alone. In the survey study we saw that consumers with different shopping practices tend to differ more in their use of certain product alternatives than in some typical ideas associated with awareness of sustainability issues. These results suggest that sustainable consumption behaviour is connected to what is going on in the realm of social practices of food consumption and the interaction with infrastructures of consumption (locales and systems of provisioning). The catering study demonstrated the thoroughly contextual nature of food consumption practices, in this case in the work place. It particularly illustrates how 'sustainable consumption' is orchestrated, experienced and learned in group- and site-specific ways, and the influential role of food professionals in enabling sustainable food consumption of consumers. A practice approach also offered an understanding of the opportunities for change and the leverage points for transitions into more sustainable food consumption in the future. For instance, both empirical studies revealed Dutch consumers' great affinity with the concept of wastefulness. Considering that food wastage, packaging and seasonality are important to such a large proportion of Dutch consumers, food providers might benefit from incorporating them more actively into their sustainable development strategies. Alternatively, the food experience strategy offered caterers the opportunity to generate positive food experiences. The canteen represents a setting in which sustainable food practices can be (re)contextualised and (re) framed, a process in which food professionals play a vital role. Attention for culinary content can be used as a strategy for improving on both food related health and sustainability, something which is now even recognised amongst nutrition scientists. Creating a better food infrastructure includes competence building amongst such actors as catering managers and cooks, and investing in the food infrastructure (i.e. kitchen and canteen facilities) of institutions and organisations.

# About the author

Elizabeth Sargant (1978, Germany) was educated at the University of Essex where she studied Ecology and Environmental Biology and Wageningen University where she studied Biology with a minor in environmental policy.

In 2005 she started working as a PhD student at the Environmental Policy Group (ENP) on a project co-funded by GaMON (gamma-onderzoek milieu, omgeving, natuur) and KSI (knowledge network on system innovations and transitions). One of the main aims of this research project was to study (perspectives on) transitions to more sustainable consumption patterns form a more practice-oriented approach, taking into account the position and role of citizen-consumers within everyday-life situations and contexts of consumption.

In 2012 she began working at Lazuur Food Community in Wageningen, a community-supported organic shop focussed on high-quality organic produce sourced as much as possible from local and regional farmers and producers, and farmers/producers who excel in realising business which are socially and ecologically innovative.

Printed in the United States
by Baker & Taylor Publisher Services